CW00322516

The Report of the Expert Group on
Animal Feedingstuffs

to the Minister of Agriculture, Fisheries and Food,
the Secretary of State for Health
and the Secretaries of State for Wales,
Scotland and Northern Ireland

London : HMSO

© Crown copyright 1992
Applications for reproduction should be made to HMSO
First published 1992

ISBN 0 11 242936 X

CONTENTS

Date **15 June 1992**

FACULTY *of*
AGRICULTURAL
and
FOOD SCIENCES
—

**Department of
Physiology
and Environmental
Science**

Dear Minister

EXPERT GROUP ON ANIMAL FEEDINGSTUFFS

I am now sending you a copy of the Report.

The remit required a wide and comprehensive review.

My colleagues and I would be pleased to discuss the content
and recommendations of the report if you wish.

School of Agriculture
Sutton Bonington
Loughborough
Leicestershire
LE12 5RD
—

Telephone
(0602) 484848
—

Telex
37346
(Uninot G)
—

Facsimile
(0509) 674384

Yours sincerely

G E Lamming

The Rt.Hon John Gummer, MP.,
Minister of Agriculture, Fisheries & Food,
Whitehall Place,
London,
SW1A 2HH

The Rt.Hon Mrs Virginia Bottomley, JP., MP.,
Secretary of State for Health,
Department of Health,
Richmond House,
79 Whitehall,
London SW1A 2NS

The Rt.Hon David Hunt, MBE., MP.,
Secretary of State for Wales,
Gwydyr House,
Whitehall,
London SW1A 2ER

The Rt.Hon Ian Lang, MP.,
Secretary of State for Scotland,
Dover House,
Whitehall,
London SW1A 2NS

The Rt.Hon Patrick Mayhew, QC, MP,
Secretary of State for Northern Ireland,
Whitehall,
London SW1A 2AZ

Professor of Agricultural Botany: W J Whittington   Professor of Animal Physiology: G E Lamming

Professor of Environmental Physics: M H Unsworth   Professor of Plant Physiology: D Grierson

CHAPTER 1

INTRODUCTION

REASON FOR REVIEW

1.1 The House of Commons Agriculture Committee's fifth report on Bovine Spongiform Encephalopathy, session1989-90 recommended "that the Government establish an expert committee to examine the whole range of animal feeds and advise on how the industries which produce them should be regulated". The Government accepted this recommendation in the response made by the Minister of Agriculture on 6 February 1991 in the House of Commons to a Parliamentary Question from Paul Marland M.P.

MEMBERSHIP

1.2 Professor G E Lamming, Professor of Animal Physiology, Nottingham University (Chairman).
Professor P C Thomas, Principal and Chief Executive, Scottish Agricultural College.
Mr C Maclean, Technical Director, Meat and Livestock Commission.
Dr E M Cooke, Deputy Directory, Public Health Laboratory Service.
Secretariat — Mrs E Owen, Miss J Powis.

TERMS OF REFERENCE

1.3 The Expert Group on Animal Feedingstuffs met for the first time on 15 February 1991 with the following terms of reference:—

"To review the existing regulatory framework covering the animal feed industry in the United Kingdom. To advise on whether any improvements are required in the mechanisms by which the responsible Departments take account of food safety requirements in regulating the industry and to report to Ministers by the end of 1991."

METHOD OF WORKING

1.4 Given the number of different regulations governing animal feedingstuffs manufacture the Group decided to examine them sectorally by subject: Bovine Spongiform Encephalopathy and other Spongiform Encephalopathies; Salmonella and other Pathogens; Medicated Feedingstuffs; Feed Composition and Marketing, and Methods of Analysis.

1.5 The Group invited comments from industry and other interested organisations, including consumer groups, by open invitation in a Press Notice issued on 6 February 1991 and from a number of organisations specifically by letter sent out on 21 February 1991. A list of the organisations which responded is at Appendix III. A number of organisations asked to meet the Group to discuss their written comments. The Group itself invited a number of individual experts to assist in its discussions.

1.6 A list of organisations and individuals who attended meetings is attached at Appendix IV. We would like to record our thanks to all those who gave their time and effort to this review.

OTHER COMMITTEES

1.7 A number of other expert committees had examined or were concurrently examining various issues which directly or indirectly relate to animal feedingstuffs. Those whose work is complete are the Working Party on Bovine Spongiform Encephalopathies; the Consultative Committee on Research into Spongiform Encephalopathies; the House of Commons Agriculture Committee Report on BSE and the Committee on the Microbiological Safety of Food. Committees which are extant are the Spongiform Encephalopathy Advisory Committee, the Veterinary Products Committee, the Interdepartmental Committee on Animal Feedingstuffs and the EC Scientific Veterinary Committee (Animal Health and

Veterinary Public Health Section). We were able to refer to their reports and in some instances have access to their deliberations. This was most helpful.

UK AND EC DEVELOPMENTS DURING COURSE OF REVIEW

1.8 EC and UK regulations are subject to frequent revision. During the course of our review the Commission of the European Communities (EC Commission) proposed harmonised controls for organisms causing zoonoses in feedingstuffs, dietetic feedingstuffs, silage agents and undesirable substances in feedingstuffs. New UK regulations were issued on bovine spongiform encephalopathy (BSE), medicated feedingstuffs, feedingstuffs marketing, heat treatment of feedingstuffs (in Northern Ireland only), and veterinary residues.

1.9 Feedingstuffs legislation is complex and fragmented. There are eleven regulations extant which are enacted under three separate Acts. We found the legislation difficult to follow and in the longer term it could with benefit be simplified. There are also nine Codes of Practice.

MAIN AREAS CONSIDERED

1.10 The main areas are considered Chapter by Chapter as shown below. We found no implications for human health in regulations concerned with pet food and zoo animals and these areas are not discussed in our report.

Bovine Spongiform Encephalopathy and other Spongiform Encephalopathies

1.11 We believe that, on present evidence, the feed bans now in place will be effective in controlling the spread of bovine spongiform encephalopathy (BSE). We have considered the risk that further species of livestock could contract spongiform encephalopathies (SEs) through feed, but have concluded that unless new evidence becomes available there is no need to supplement the feed controls now in place. However additional areas of research should be examined and surveillance of livestock should be heightened to identify early the development of SEs, or other feed-borne infections.

Salmonella and other Pathogens

1.12 We recognise that feedingstuffs can be a source of new salmonella serotypes, but we are unable to determine the extent to which they contribute to animal carriage and disease. Further investigations throughout the feed, animal and human food chain using standardised scientific and statistical methods are necessary.

Medicated Feedingstuffs

1.13 In general we are satisfied with the controls on medicated feedingstuffs. Some fish farmers using medicinal additives are not required to register with the Royal Pharmaceutical Society of Great Britain or in Northern Ireland the Department of Health and Social Services. We consider all feed manufacturers using medicinal additives should be subject to control and inspection.

Feed Composition and Marketing

1.14 We identified some deficiencies in primary powers, for example, for routine inspections of raw materials and feedingstuffs at ports of entry to the UK and on some farms, and we make recommendations to rectify them.

Methods of Analysis

1.15 There were no substantial problems in this sector. We noted however that new methods for determining certain feed constituents are developed too slowly and we believe that this should be corrected.

MAIN RECOMMENDATIONS

Surveillance arrangements

1.16 We regard feedingstuffs as an integral part of the feed/farm animal/human food chain for the purposes of ensuring food safety and our comments and recommendations contained in the report reflect this view.

Legislation

1.17 We found that the feed industry is regulated by extensive controls, which taken together, are in most areas effective in safeguarding food for human consumption. However there are gaps in legislation and in its enforcement.

Animal Feedingstuffs Advisory Committee

1.18 There is, in our view, a need to streamline and extend the process by which expert advice is coordinated and presented to Ministers and officials. We recommend that an independent Animal Feedingstuffs Advisory Committee be established to take an overview of all feedingstuffs issues. This Committee's composition and terms of reference are proposed in detail at Chapter 7 but is also referred to at other stages earlier in the Report.

CHAPTER 2

BOVINE SPONGIFORM ENCEPHALOPATHY AND OTHER SPONGIFORM ENCEPHALOPATHIES

INTRODUCTION

2.1 In examining this area of legislation, the Expert Group has taken into account the work already carried out by independent committees set up in the UK following the emergence of BSE:—

(i) *Working Party on Bovine Spongiform Encephalopathy* (Southwood Committee) — set up in May 1988 under the chairmanship of Sir Richard Southwood FRS to examine the implications of Bovine Spongiform Encephalopathy (BSE) in relation to both animal health and any possible human health hazards, and to advise the Government on any necessary measures.

(ii) *Consultative Committee on Research into Spongiform Encephalopathies* (the first Tyrrell Committee) — set up in 1989 on the recommendation of the Southwood Committee to advise the Government on research on transmissible spongiform encephalopathies (SEs). Its chairman was Dr D A J Tyrrell, FRS. It was superseded in April 1990 by the Spongiform Encephalopathy Advisory Committee, which is under the same chairmanship.

(iii) *Spongiform Encephalopathy Advisory Committee* (the second Tyrrell Committee, also known as SEAC). Its remit is to advise the Government on all matters related to SEs and to overview research work. We have received oral evidence from members of this committee.

We have also taken into account the views of the House of Commons Agriculture Committee which reported on BSE in July 1990, and the EC Scientific Veterinary Committee Public Health and Animal Health Sections, which have also considered SEs.

BACKGROUND TO THE DISEASE

2.2 BSE is a progressive, ultimately fatal neurological disorder of cattle, first confirmed in November 1986 by the Central Veterinary Laboratory, Weybridge. It has a long incubation period, with most cases occurring between the ages of 3 and 5 years. Up to 5 June 1992, 58,880 cases of the disease had been confirmed in 18,527 herds in Great Britain, 92.5% in dairy herds and 7.5% in beef herds. The number of confirmed cases is still rising. Epidemiological studies indicate that BSE was caused initially by feeding cattle rations containing meat and bone meal which was prepared from animal material including tissues infected with scrapie, a disease of sheep endemic in several countries including the UK for over 250 years. Subsequently the epidemic was amplified by feeding meat and bone meal containing BSE-infected cattle material to other cattle.

2.3 Two of the major predisposing factors to the occurrence of BSE were:—

(a) a wide geographic distribution of endemic scrapie.

(b) significant growth in the sheep population to a level where sheep material reached about 15% (by weight) of all material rendered in Great Britain. Bovine material constituted 45% of the total.

The epidemiological evidence suggests that the events triggering the outbreak of BSE occurred in the winter of 1981/82 and probably included changes in the rendering process. It is hypothesised that a reduction in the use of hydrocarbon solvents to extract tallow during rendering, and an associated reduction in the application of moist heat in the low lipid environment of the solvent extracted material, combined to allow sufficient infective dose of the scrapie agent to survive the rendering process and cause infection in cattle. During the earlier epidemiological studies it was thought that the introduction of continuous rendering systems in the 1970s and 1980s may have been a factor in the emergence of BSE (see Southwood Report Para 4.2.8). However, further studies have indicated that the time-scale of the introduction of continuous rendering is not consistent with the pattern of the development of the epidemic (see Annex 2.4 for details of rendering systems).

2.4 Monitoring of BSE cases appears to confirm that feed was the source of infection. To date only one case of BSE has been confirmed in an offspring of a BSE positive dam, born after the feeding of ruminant protein to ruminants was banned in July 1988, where there is no evidence that the offspring had received ruminant protein. Also the incidence of BSE in offspring of confirmed cases is the same as the incidence in cattle whose dams are unaffected. The indications are therefore that maternal transmission, if it is occurring, is at a very low level and should not play a significant part in the future progression of the disease. Provided there is no direct cattle to cattle transmission, and the ruminant feed ban is effective, BSE should eventually be eliminated.

2.5 From October 1991, there is evidence of a decline in the rate at which home-bred herds are experiencing BSE cases for the first time, and recently there has been a decline in the incidence in 2-year-old animals, consistent with the expected effect of the ruminant feed ban introduced in July 1988. Finally, the within-herd incidence of BSE has changed little since 1988, whereas it would have been expected to increase if direct cattle to cattle transmission was important. These early signs lend confidence to the conclusion that the measures introduced to control the disease are proving effective.

Other Countries 2.6 Up to 30 April 1992, cases of BSE had been confirmed in the Irish Republic (49 cases), Switzerland (12 cases), France (5 cases), Oman (2 cases) and the Falkland Islands (1 case). The cases in Oman and the Falkland Islands occurred in animals imported from Great Britain, the Swiss and French cases in indigenous animals. The sources of infection are not yet clear but it is known that in the French and Swiss cases, meat and bone meal cannot be ruled out.

CURRENT CONTROLS ON BSE 2.7 Statutory measures for the control of BSE, including provisions for feed, are contained in the BSE Order 1991 which came into force on 6 November 1991. This Order revoked the BSE (No2) Order 1988 as amended; which in turn replaced the first BSE Order, introduced earlier that year.

2.8. The main measures with implications for feed are as follows:—

1. *Destruction of carcases of cattle affected with BSE* The disease was made notifiable in Great Britain in June 1988. Following interim advice from the Southwood Committee, arrangements came into force on 8 August 1988 for the compulsory slaughter of cattle with clinical signs of BSE, with disposal of carcases by incineration or burial. This was to ensure that carcases of suspect animals would not enter the human food chain, either directly or through animal feed. The same arrangements were introduced in Northern Ireland in November 1988.

2. *Ban on Feeding Ruminant Protein to Ruminants* In the light of the epidemiological studies, the use of ruminant-based protein in ruminant rations was temporarily banned in Great Britain from 18 July 1988. The ban was introduced in Northern Ireland on 11 January 1989. Both bans were extended indefinitely by the BSE (No2) Order 1988 on the advice of the Southwood Committee, because there was uncertainty that any of the existing rendering systems would inactivate the agent. This ban has been maintained in the BSE Order 1991.

3. *Ban on the inclusion of specified bovine offals in animal rations* This ban was introduced in September 1990 on the advice of the second Tyrrell Committee. The specified bovine offals are those bovine tissues which, by analogy with scrapie, were thought most likely to contain the BSE agent. They are the brain, spinal cord, thymus, tonsils, spleen and intestines of bovine animals over 6 months of age. The ban applies to feed for any animal or bird, including poultry. The specified offals had already been prohibited for human consumption under the Bovine Offal (Prohibition) Regulations 1989, and this prohibition was extended to animals after experimental (parenteral) transmission of BSE to a pig. (This experiment is described more fully in Para 2.24). It should be noted, however, that an SE has never been identified in the pig under field conditions anywhere in the world, although pigs must have been exposed to the agent through consumption of feed containing infected ruminant protein.

4. *Milk from cattle with suspected BSE* The sale or use of milk from suspect animals for human or animal consumption was banned from 30 December 1988 in Great Britain on the recommendation of the Southwood Committee. An exception is the use of the affected cow's milk to feed its own calf, necessary occasionally for practical and welfare reasons. This was a precautionary measure, as milk has never shown infectivity nor been implicated in the transmission of SEs including BSE. Similar, non-statutory, arrangements are in place in Northern Ireland.

EFFECTIVENESS OF CURRENT CONTROLS

2.9 We are satisfied that the arrangements for the compulsory slaughter and total destruction of the carcases of BSE suspects are effective and that such carcases present no risk via animal feed. Therefore, in our examination of current control measures, we concentrated on the two measures most directly concerning feed — the ruminant protein ban and the specified bovine offals ban:—

1). *The ban on the feeding of ruminant protein to ruminants* This measure, if properly enforced, will be effective in preventing the exposure of cattle to BSE infection via feed and in protecting all other ruminant species from exposure to food-borne sources of ruminant SE agents. The inclusion of both ovine and bovine sources of protein in the ban will prevent the infection of cattle with scrapie from sheep, and will stop the recycling via feedingstuffs of SEs in cattle and sheep. It will also protect exotic Bovidae, some of which have succumbed to SEs in British zoos, and deer which are also susceptible to an SE (chronic wasting disease) though no such disease has occurred in the UK. Although the ban came into force in July 1988 it was expected that it could require up to six months for all ruminant rations, manufactured before the ban, and containing ruminant protein, to be cleared from the feed chain. Current epidemiological evidence suggests that the effective lag in clearing these feed sources has been about three months.

2). *The ban on the feeding of specified bovine offals or protein derived from them to all animals* We regard this as an important precautionary measure, because it removes tissues which may contain BSE infectivity from the feed chain, including those from animals not exhibiting clinical signs of disease. The BSE agent has not so far been identified in any bovine tissue except the brain. This is only one of the specified bovine offals, reinforcing confidence that no potentially BSE-infected tissues are fed to animals or birds.

2.10 We examined in detail two additional aspects of the ban on the use of specified bovine offals.

Tallow

2.11 Currently, tallow derived from specified bovine offals can be incorporated into animal feedingstuffs, as the ban on use applies only to the specified offals themselves or protein derived from them. The Southwood Committee had concluded that there was no evidence that tallow had been a source of BSE infection (Southwood Report Para 4.2.3). Their conclusions were based on two factors. Firstly, the BSE agent is proteinaceous, and therefore will tend to partition more with the protein fraction (meat and bone meal) rather than with the lipids of tallow. (In addition, all feed-grade tallow, in order to conform to

British Standard Specification (BS 3919:1987), is subjected to vacuum filtration, which reduces the protein content to approximately 0.1%. Any contaminating infectivity in tallow derived from specified offals would in turn be substantially reduced). Secondly, the geographical distribution of BSE cases closely reflects regional differences in the processes used for the production of meat and bone meal, the amount of reprocessed greaves included in the raw material, and the variation in the market share of compounders who use meat and bone meal in their products. The disease distribution does not reflect the distribution pattern for tallow, which unlike meat and bone meal, is often incorporated into animal feed long distances from the area where it was produced. **On the basis of this evidence, we are satisfied with the current controls.**

Exemption of specified offals from cattle under 6 months of age

2.12 We considered whether calves under 6 months of age could be sub-clinically infected with BSE, and their offals represent a potential hazard to animal health if incorporated into feed. Evidence was presented to us by MAFF and the Tyrrell Committee that natural scrapie infection has rarely been detected in sheep under 10 months old and the youngest case of BSE occurred at 22 months of age. Using the evidence of the research from scrapie, the level of infectivity, if present in any of the offals of young animals, would be extremely low and probably undetectable. In the absence of evidence of vertical or horizontal transmission of BSE, and provided the feed bans were effective, calves would not have been exposed to infection through BSE- or scrapie-contaminated meat and bone meal (allowing for the time lag in using existing stocks) since December 1988 in Great Britain, and June 1989 in Northern Ireland. **Given these factors, we are content that the tissues from calves under 6 months of age should continue to be exempt from the specified offals ban.**

2.13 The effectiveness of the current control programme for BSE is difficult to determine at this stage, because the long incubation period dictates that any decline in the incidence of the disease would not start before late 1992, at the earliest. However, as indicated in paras 2.4 and 2.5 above, the evidence is that spread of the disease from animal to animal is unlikely to be important. **We recognise the importance of both feed bans in the control of BSE, and in minimising the exposure of other species to the agent and *recommend* that they be maintained.**

ENFORCEMENT OF CURRENT CONTROLS
Slaughterhouses

2.14 Current arrangements for enforcement of controls are as follows:—

1) The local authorities enforce the Bovine Offal (Prohibition) Regulations. Meat inspectors employed by the local authorities ensure that extraction of specified bovine offals from the carcase is carried out satisfactorily, and that they are separated from other waste material. In Great Britain under the Regulations, the specified bovine offals must, except under certain specific circumstances, be either 'sterilised' or stained before being moved, under cover of a local authority permit, to a specified destination — usually a renderer. In Northern Ireland, officials of the Department of Agriculture for Northern Ireland (DANI) enforce the legislation, which parallels the provisions in Great Britain. However, in Northern Ireland, the staining and sterilisation of specified bovine offals is not required if being moved under licence to a renderer.

Knackers

2) Knackers are licensed and visited periodically by the local authorities, and are now subject to monthly visits by MAFF. In addition to other duties, these visits help to ensure that the specified bovine offals are kept separate from other waste and are dispatched to renderers, again under a permit issued by the local authority. MAFF also makes monthly visits to hunt kennels (another recipient of fallen stock) to ensure compliance with the ban on the use of specified bovine offals for feeding to animals.

Renderers

3) Renderers must process specified bovine offal separately from material intended for use in animal feedingstuffs. In Great Britain, under the BSE Order 1991 meat and bone meal derived from specified bovine offals may only be moved from renderers under MAFF licence, normally for burial in land-fill sites or for incineration. MAFF visits the plants every 2 months to ensure that this meat and bone meal does not enter the animal feed chain.

It has powers to check that, once dispatched, specified material reaches its approved destination. In consultation with the industry, MAFF is preparing a code of practice for use in plants which render both specified bovine offals and other animal waste. Once adopted this code should further ensure the separation between the different types of material. With regard to the ruminant protein ban, most renderers process material from cattle, sheep and pigs together, since it would be impractical for them to do otherwise. Only in the case of poultry materials are there dedicated plants. In these circumstances, the only meat and bone meal which could be fed to ruminants would be derived from poultry. DANI enforces the equivalent legislation in Northern Ireland.

Feed compounders

4) In Great Britain, local authorities are responsible for the enforcement of the BSE Order (essentially the ruminant protein ban) in the feed compounding industry. DANI is responsible for the enforcement of the equivalent legislation in Northern Ireland. At present there is no method of analysis available for detecting the presence of ruminant protein in ruminant rations. Enforcement is therefore difficult.

Farmers

5) Local authorities in Great Britain, and DANI in Northern Ireland, are responsible for enforcing the legislation on-farm, where the enforcement difficulties are the same as those presented at feed compounders.

2.15 The evidence suggests that in the majority of cases, the controls are working, despite the fact that the ruminant protein ban and the specified bovine offals ban are to a considerable extent dependent on self-regulation by the industry. However, there are indications that some cattle may have had access to ruminant meat and bone meal since the 1988 ban and it is thought that this may account for most of the 69 confirmed cases of BSE in cattle which, up to 5 June 1992, have been born after the ruminant feed ban was introduced. The majority were born shortly after the statutory intervention. This number is small compared with the earlier incidence, but it suggests that the integrity of the ban is not complete. Although there are signs that this may be a diminishing problem, **we recommend that the incidence of BSE in cattle born after the ruminant protein ban is monitored carefully. We also welcome the development of the tests for the detection of ruminant protein in meat and bone meal and compound feedingstuffs [see Para 6.2] which, when they become available for enforcement purposes, will provide an additional safeguard.**

EC DEVELOPMENTS:
Directive on the disposal of
animal waste (*90/667EEC*)

2.16 This directive came into force at the end of 1991, and deals, primarily, with the rules governing the operation of the rendering industry. Its primary objective is to minimise the risk of salmonella infection in livestock through the consumption of contaminated meat and bone meal, and should provide for free movement of meat and bone meal within the Community provided it is processed in accordance with the terms of the Directive. **Theoretically, the feed controls for BSE in place in the UK would be in contravention of this Directive, as they apply to imported, as well as home-produced, material. However, the Directive contains a provision which, by permitting the retention of national veterinary legislation applicable to the eradication and control of certain diseases, allows the retention of the current UK feed bans. We recommend that this provision be invoked.**

EFFECTIVENESS OF
RENDERING

2.17 Despite the fact that experts were satisfied that the transmission of BSE would be controlled by the ruminant protein ban, attention was also directed to methods to destroy the agent in the rendering process. The Southwood Committee recommended studies on the temperature and treatment conditions required to inactivate the agent and this was supported by the first Tyrrell Committee. Finally, the House of Commons Agriculture Committee recommended that the Expert Group investigate the feasibility of developing a rendering process capable of destroying the BSE agent, and consider whether this should be made a research priority. The Governments' response to the Agriculture Committee report indicated that such research was already in hand. MAFF, the EC, the European

Renderers Association, the Agriculture and Food Research Council, and the UK rendering industry are collaborating on a series of experiments to evaluate the effectiveness of the majority of UK and Community rendering systems in inactivating the agents which cause BSE and scrapie. The results of these studies using BSE-infected material will not be available for some time, while the analogous scrapie experiment still requires more scrapie-infected material to be collected before it can start. Two further inactivation studies are in progress: one investigating the susceptibility of SE agents to a range of chemical and physical treatments, and the other their susceptibility in laboratory facsimiles of rendering practices.

2.18 In the Southwood Report, the possibility was raised of lifting the feed bans, if a rendering process capable of destroying the agent could be developed. This possibility is also recognised in the EC Directive 90/667 on the disposal and processing of animal waste. This contains a provision enabling amendments to be made to allow rendering conditions for high risk material to be modified if scientific evidence is produced in support of such a change. If results of the inactivation studies show clearly that certain systems will destroy the agent it would be possible to use these as evidence to seek a change in allowable processing conditions. We are concerned about this possibility. There is doubt that rendering plants could consistently meet the necessarily demanding criteria for processing unless the acceptable time/temperature conditions were set with a very high safety margin. With such a margin there is a real probability that the resulting protein would be denatured, making the process unattractive commercially. We are aware of the disposal problems associated with meat and bone meal derived from specified bovine offals. Nevertheless, given the concerns cited above, **we recommend that the feed bans be retained even after the results of the inactivation study become available, unless the results provide unequivocal information on the inactivation of the scrapie/BSE agents, and the necessary conditions can be consistently achieved by the rendering industry. In our view, the feed bans are, at the present state of our knowledge, the only certain safeguard in preventing disease transmission via feed in animals.**

CURRENT RISKS OF INFECTION OF OTHER SPECIES BY SE AGENTS VIA FEED

2.19 We have attempted to assess the likely risks of livestock other than ruminants succumbing to an SE in the UK via feed under current circumstances. The following is a synopsis of evidence reviewed:—

Scrapie

Scrapie is the only SE known to occur as an endemic infection of its natural hosts — sheep, and probably goats. Scrapie infection is probably maintained in the sheep population because it is transmitted maternally and horizontally. It provided the most likely source of BSE and probably that for transmissible mink encephalopathy.

Transmissible Mink Encephalopathy (TME)

This disease occurs in ranch-reared mink and is thought to be due to exposure to infection through the feeding of scrapie-infected unprocessed sheep offal. It is possible that transmission may also be by direct injection into wounds as a result of fighting between kits. There is some unsubstantiated evidence that TME may have occurred through feeding untreated material from fallen and sick cattle, in an incident where it was reported that no sheep material had been fed. Morbidity of some outbreaks has approached 100% of adults, probably due to exposure to high levels of infectivity in the sheep offal due to the absence of any heat treatment. The rendering of offal for domestic animal use reduces the risk of these circumstances being repeated elsewhere. TME is a rare condition, with only approximately 20 outbreaks in the last 40 years. None has occurred in the UK.

Kuru

Kuru is a human SE that is associated with 'feeding practices'. Its origin is unknown but may have been infection from an isolated case of another rare human condition — Creutzfeldt Jakob Disease (CJD). The disease was 'passaged' within a human population in Papua New Guinea through cannibalism practised by relatives during mourning rites. Since the cessation of cannibalism, kuru has become a rare disease.

Chronic Wasting Disease (CWD)

This SE occurs in wildlife facilities in Colorado and Wyoming in the USA in Rocky Mountain elk, mule deer, black-tailed deer and progeny of the crosses of mule deer and white-tailed deer. The disease is not believed to be the result of exposure to a foodborne agent. It is maintained in captive populations probably by horizontal and possible maternal transmission. However, it is also possible that infection may have come from a natural reservoir of infection in free range deer. Some cases of CWD have been found in wild deer in the locality of the wildlife parks in the USA.

Spongiform Encephalopathies in Exotic Bovidae

In the UK, a small number of incidents of SE has been confirmed in five species of exotic Bovidae*. Two species are referred to in the Southwood Report (Para 4.2.4). These animals were fed the same type of concentrated feeds as cattle and could therefore have been exposed to the infective agent that led to BSE. There is evidence that the exotic Bovidae may be more susceptible to infection than domestic species; the age of onset was low in some cases indicating a shorter incubation period. Maternal transmission may have occurred in the case of a calf born to an infected greater kudu. The risks of SE infection in exotic Bovidae has now been removed through the introduction of the ruminant protein ban.

Spongiform Encephalopathies in Felidae

25 cases of feline spongiform encephalopathy (FSE) have been confirmed in the UK between May 1990 and June 1992. FSE is almost certainly a new disease because its characteristic clinical signs would have ensured its recognition if it had occurred earlier. Furthermore, histopathological studies of preserved cat brains in archive failed to find signs of SE. The timing of first occurrence, in 1990, suggests that FSE is more likely to be a consequence of BSE than of scrapie. Also in the UK, there has been one case of a spongiform encephalopathy in a captive puma (*Panthera concolour, Felis concolour*), which may have been infected via feed. The risk of transmission of BSE to pumas and domestic cats has now been removed by the specified bovine offals ban.

Transmission between species: the species barrier

2.21 The transmission of SE agents between species is dependent upon the effective dose given, which is a function of three factors:—

(a) the amount of agent administered;

(b) the route of administration; in studies in the mouse, for example, the efficiency of infection with scrapie by the oral route was 100,000 times less than by the intracerebral route;

(c) the species barrier effect, which varies with the strain of agent and the donor/recipient species.

The species barrier is nearly always the major factor limiting the interspecies transmission but it can usually be overcome by parenteral exposure to very high titres of infectivity. For example, BSE transfer to pigs and marmosets has been achieved through use of massive

---

*Nyala *(Tragelaphus angasi)*, gemsbok *(Oryx gazella)*, Arabian oryx *(Oryx leucoryx)*, greater kudu *(Tragelaphus strepsiceros)*, and eland *(Taurotragus oryx)*. They are from the same Family (Bovidae) as cattle and sheep.

doses of infective material administered by a combination of parenteral routes, including intracerebrally. Once the species barrier has been overcome, however, transmission within the species is possible and the incubation period length is usually reduced.

TME

TME is an example of interspecies transmission of an SE but with no natural means of maintaining the infection in the species. The mink is a 'dead-end' host. TME is caused by exposure to large titres of infectivity through unprocessed, infected animal protein.

Kuru

Kuru is an example of a disease which became endemic through recycling infection within a species. Informed opinion suggests kuru may have started with a case of CJD. Thus no species barrier was involved in either initiating or sustaining the epidemic.

BSE

Scrapie was probably the source of BSE, so the primary disease occurred because the species barrier from sheep to cattle was breached. However, there is evidence that before the first clinical cases of BSE were recognised in 1986, infected cattle material was being recycled in meat and bone meal and fed to other cattle. This produced the equivalent of serial 'passage' and would have favoured the selection of cattle-adapted strains of agent and allowed multiplication of infectivity during each passage. Most importantly, there would no longer have been a species barrier. This would give BSE strains the selective advantage over scrapie strains. This caused an upsurge of BSE cases starting mid-1989. The greatest risk of another BSE-like epidemic in domestic species would therefore be from the same combination of inter-species exposure to infection and intra-species recycling of infection.

2.22 On the basis of this information and logic, the various feeding practices using animal protein can be ranked according to risk; starting with the highest:—

a) a combination of interspecies exposure and intra-species recycling of infection (see above);

b) the recycling of infection within species;

c) the inter-species exposure of infection.

We therefore explored the risks to sections of the animal population posed by present feeding practices.

Current possibilities of infection of UK livestock by SE agents

2.23 There are two reservoirs of SE agents in livestock in the UK: the natural reservoir of scrapie in sheep and that which was created as BSE in cattle by the use of contaminated feedingstuffs. The extent of the current exposure of livestock to the known reservoirs of infection is outlined below. Those situations not covered by existing bans are underlined:—

Known reservoir	Preventative measures	Species exposed via feed
SHEEP	- - - - - - / RPB / - - - - - -➤	Cattle & other ruminants
SHEEP	- - - - - - - - - - - - - - - - -➤	Pigs
SHEEP	- - - - - - - - - - - - - - - - -➤	Poultry
SHEEP	- - - - - - - - - - - - - - - - -➤	Fish
CATTLE	- - - - - - / RPB / - - - - - -➤	Sheep & other ruminants
CATTLE	- - - - - -/ SBOB / - - - - -➤	Pigs
CATTLE	- - - - - -/ SBOB / - - - - -➤	Poultry
CATTLE	- - - - - - - - - - - - - - - - -➤	Fish

RPB — ruminant protein ban   SBOB — specified bovine offals ban

11

Therefore, the livestock currently potentially exposed to SE agents are pigs and poultry to the scrapie agent, and fish to both the scrapie and BSE agents. The risk of any of these livestock developing an SE under current circumstances appears small.

Pigs

2.24 During the time that cattle received a feed exposure that led to a major epidemic of BSE, pigs would have received an even greater exposure, because their diet contained higher inclusion rates of meat and bone meal, without any evidence of cases of SE. From the size and age structure of the breeding sow population, it can be calculated that if the effective exposure had been the same for pigs as it was for cattle, about 1,000 cases of 'porcine spongiform encephalopathy' would have been seen by now. Experimental transmission of BSE to pigs has indicated the symptoms to be expected in the naturally occurring disease. If it had occurred, there can be little doubt that it would have been recognised. The experimental transmission to pigs was achieved by inoculating massive doses directly into the brain, blood stream and peritoneal cavity. None of the pigs orally exposed to similar, repeated, large doses has succumbed though the experiment is incomplete. Furthermore, the specified offals ban has now removed BSE as a source of infection.

2.25 It is possible that new scrapie strains might appear in sheep that could be more pathogenic for pigs than any existing strains. There are at least two possible sources of such strains. One source could be a mutation of an existing scrapie strain. Another might result from the rare introduction via feed (before July 1988) into a small number of sheep, of a BSE agent derived from infected cattle offals. However, this possibility seems unlikely because:—

a) the extent of concentrate feeding to sheep kept for breeding is small compared to dairy cattle;

b) cattle-adapted BSE strains would be at a selective disadvantage in sheep because they would have to cross a species barrier.

2.26 If the new scrapie strain in sheep from either source (mutation or feed) developed it could theoretically be more pathogenic for pigs than any strain hitherto. However, selection of this mutant would be necessary for it to become prevalent in sheep. The ruminant protein ban would prevent artificial selection; and natural selection in sheep would depend upon there being a positive selection pressure. Therefore, this theoretical risk is maximised if three chance events occur:—

(a) the occurrence of a mutant strain in the first place — a relatively likely event;

(b) its natural selection and spread through a sizeable part of the sheep population — which is unlikely;

(c) and a greater pathogenicity for pigs than established strains.

If such a combination of events were to occur, the recycling of the new strain in pigs could certainly favour its selection, but its initial selection in sheep and transmission to pigs is unlikely.

2.27 Our conclusion is that pigs are less susceptible to UK strains of scrapie and BSE than cattle. The specified offals ban eliminates BSE as a source of risk and this ban should stay in force. There are still theoretically risks from sheep scrapie; they are remote, but nevertheless we later recommend increased surveillance and research.

Poultry

2.28 The gene associated with susceptibility to the encephalopathies in mammalian species has also been found in chickens, but with substantial differences as compared with the mammalian genes. There is no evidence of a naturally occurring SE in poultry, but the relatively short commercial life span of birds may limit the value of this negative evidence. In the late 1980s, several cases of SE occurred in ostriches (long-lived birds) in Germany.

Although brain histology revealed lesions of a scrapie-like spongiform encephalopathy, no unfixed brain material was available to look for the modified (fibrillar) form of PrP protein, another pathological sign. Only very small amounts of formalin-fixed brain were obtained for current transmission studies in laboratory rodents which were set up recently in Germany. The ostriches were fed commercial poultry feed, vegetable matter and bovine material from emergency slaughterhouses, but there is insufficient information to indicate the putative source of infection.

2.29 Limited attempts to transmit SE diseases to chickens and other domestic and wild avian species have been unsuccessful. Studies were initiated by MAFF in the summer of 1990 in which chickens were challenged with brain from BSE affected cows by parenteral or oral exposure. The intention is to investigate multiplication of the agent in various tissues as well as the occurrence of clinical disease. Results will not be available for at least another year. However, at the present time, the available evidence suggests that susceptibility of chickens to SEs may be low. A large species barrier would be expected to limit the transmission of SE agents from mammals to birds.

2.30 **It is difficult to assess the risks to chickens without further information from current transmission experiments, and we therefore recommend that experimental work in this area continues, and the situation reappraised in the light of the results.**

Fish

2.31 There has been a rapid growth in fish farming in recent years. Fish are fed diets containing both fish and animal protein. Since the specified offals ban does not apply to fish, it is possible that they could be exposed to BSE as well as scrapie. There is no information available about the occurrence of SEs in fish or the experimental susceptibility of fish to these agents. There are no experimental studies in progress or being planned. However, presence of a PrP gene, apparently an essential requirement for species susceptibility, has been found in a wide range of vertebrates. It is therefore possible that it occurs in fish, and therefore fish have the theoretical capacity to develop SEs. However, there are other factors which indicate that the risk of infection is unlikely to be significant:— First, there are indications that a large number of fish farmers are no longer feeding rations containing any land animal protein, following the emergence of BSE. Secondly, a large species barrier would again be expected to limit the transmission of SE agents from mammals to fish. Thirdly, accepting the evidence that maternal transmission will not be a prominent feature of BSE the risks from feeding meat and bone meal from specified bovine offals (assuming that this occurs) would now be low and would decrease very rapidly once the BSE epidemic has passed its peak. The transmission of infection from sheep to fish, although remote, is still a possibility.

Risks from own-species recycling

2.32 Based on our current information, the possibility of clinical disease caused by an SE agent in each of the three main non-ruminant domestic livestock species used for food — pigs, poultry and fish — is remote. It is, however, possible that there could be inapparent infection in these species. In this case, artificial recycling of the infection within the species via feed could increase the likelihood of the emergence of a pathogenic strain of the agent, as was the case with kuru and BSE, because there would be no species barrier. Own-species recycling has already occurred in all three species but no SE has resulted to date. Fish-farming, in which own-species recycling could occur, is largely confined in the UK to trout and salmon, both of which are salmonids. As part of preventive medicine programmes, fish farmers are advised not to feed any salmonid material to their fish, and it is believed that this advice is largely observed. In response to public concern following the problem of salmonella in eggs, the poultry industry has reduced the practice of feeding poultry material back to poultry. However, it is still a significant practice: 35% of the total amount of meat and bone meal used in the UK, part of which is derived from poultry waste, is incorporated into poultry feed. The other major recipient of own-species material via feed is the pig, currently accounting for 35% of total meat and bone meal usage, a significant proportion of which is derived from pig material. The pig has traditionally been fed large quantities of meat and bone meal, including own-species material. There is evidence to suggest that pigs are not naturally susceptible to SEs (see para 2.24) and they have a long history of exposure to scrapie without any succumbing. Their traditional role as scavengers has ensured their exposure to own-species material over a long period of time, with no apparent ill-effects. Taking into account the remoteness of the risk of infection by an SE agent, and the valuable function that the pig serves as an outlet for this waste material, we consider that there is no justification at this stage for prohibiting the feeding of own-species material to pigs. However, the pig presents the greatest theoretical risk of infection by an SE agent, because it is exposed to scrapie through meat and bone meal derived from sheep material, and to own-species material in the meat and bone meal. While poultry are exposed to the same feeding practices, unlike pigs, they have so far failed to demonstrate their susceptibility to SE agents under experimental conditions.

Surveillance

2.33 The development of BSE has heightened awareness among the veterinary profession of SEs. In addition, as stated earlier, if an SE in the pig were to develop under natural conditions, it would be readily recognisable. **While acknowledging that, with the current BSE-related feed bans in operation, the risk of a recurrence of a BSE-type situation is low, we recommend that surveillance for SEs in farm animals be heightened, in order to identify rapidly the development of disease in other species, should it occur.**

Research

2.34 At present, the absence of a diagnostic test for SEs in the live animal means that it can only be identified once it has reached the clinical stage, hampering the early identification of infection and rapid introduction of any necessary control measures. **In regard to research into the early detection of SEs in asymptomatic animals, we are aware that research was already in progress when the first Tyrrell Committee reported, and they gave it the highest (3 star) priority. We recommend that research on the early detection of SEs continues at this highest level of priority.**

2.35 Currently the oral transmission experiments on pigs involve exposure to BSE. However, while the control measures in place for BSE should eventually lead to the eradication of the disease, the reservoir of infection from scrapie will remain. **We therefore recommend that, as a matter of priority, consideration be given to extending the pig oral challenge experiments to include scrapie-infected material.** Earlier in the chapter we have emphasised the importance of the species barrier in limiting the transmission of SEs between species. **We therefore recommend that Government and appropriate Committees also consider support for research and development on the species barrier as it relates to SEs and specifically on the nature of the <u>PrP</u> gene in the pig.**

2.36 We note that scrapie will become a notifiable disease in the EC by the end of 1992 and this will provide Governments with information on the incidence of the disease. **We recommend that when the results of the research on the species barrier as it relates to SEs and the information as a result of notification of scrapie are available, the UK Government should invite the appropriate committees to examine methods of reducing the reservoir of scrapie.**

ANNEX 2.1

LEGISLATION: BOVINE SPONGIFORM ENCEPHALOPATHY

ANIMAL HEALTH ACT 1981

2.1.1. The Animal Health Act 1981 provides the enabling powers for the statutory provisions for the control of the disease, including feed measures, contained in the BSE Order 1991.

THE BSE ORDER 1991 (SI NO 2246)

2.1.2. This Order came into force on 6 November 1991, and revoked the BSE(No2) Order 1988, as amended. Among other measures, it provides for the compulsory notification of BSE and for the slaughter of and payment of compensation for cattle in which BSE is suspected, with disposal of their carcases at the expense of the Ministry. It prohibits the use of ruminant protein in ruminant rations, and the use of any specified bovine offal or protein derived from it in feedingstuffs for any kind of mammal except man, any kind of non-mammalian four-footed beast and any bird. Unprocessed specified bovine offal may only be moved under licence under the Bovine Offal (Prohibition) Regulations 1989 and the Bovine Offal (Prohibition) (Scotland) Regulations 1990 (which deal with food potentially for human consumption). The same provisions apply under the BSE Order 1991 to the protein derived from specified bovine offals, and to the export of specified bovine offal and its protein to EC member states. It also prohibits the use of milk from affected or suspected cattle for both human and animal consumption, with the exception of the feeding of the cow's own calf. The export to third countries of specified bovine offal and its protein is controlled by Department of Trade and Industry legislation (The Export of Goods (Control) (Amendment No7) Order 1991), which is applicable throughout the UK.

NORTHERN IRELAND

2.1.3. The Department of Agriculture for Northern Ireland introduced measures effecting similar arrangements and controls on suspect and confirmed BSE infected animals, their disposal and compensation. The controls governing the disposal of specified bovine offals and use of ruminant protein generally also parallel those in Great Britain. The relevant principle legislative powers are contained in the Diseases of Animals Order (Northern Ireland) 1981 (as amended) and related subordinate legislation consists of the BSE Order 1988 (Northern Ireland) (SR No 422), the Diseases of Animals (Modification) Order (NI) 1990 (SR No 135) and the Diseases of Animals (Feedingstuffs) Order (NI) 1992 (SR No 215). The Bovine Offal (Prohibition) Regulations (NI) 1989 parallels the Great Britain legislation.

ANNEX 2.2

SPONGIFORM ENCEPHALOPATHY ADVISORY COMMITTEE

This Committee was established in April 1990, superseding the Consultative Committee on Research into Spongiform Encephalopathies. Its terms of reference are:—

"To advise the Ministry of Agriculture, Fisheries and Food and the Department of Health on matters relating to spongiform encephalopathies."

Members

Chairman

Dr D A J Tyrrell	Former Head of the Medical Research Council Common Cold Unit, Porton Down, Salisbury

Professor I Allen	Neuropathologist, Royal Victoria Hospital, Belfast
Professor R M Barlow	Former Veterinary Pathologist at the Royal Veterinary College, University of London
Professor F Brown	Former Deputy Director (Scientific) of the Animal Virus Research Institute, Pirbright (now known as the Institute of Animal Health)
Dr R Kimberlin	An independent consultant and former Head of the Medical Research Council/Agricultural and Food Research Council Neuropathogenesis Unit, Edinburgh
Mr D B Pepper	Veterinary Surgeon
Dr W A Watson	Former Director of the Ministry of Agriculture, Fisheries and Food (MAFF), Central Veterinary Laboratory, Weybridge
Dr R G Will	Consultant Neurologist at the Western General Hospital, Edinburgh

Observers

Mr R Bradley	MAFF Central Veterinary Laboratory, Weybridge
Dr A Wight	Department of Health

Research Council representatives

Dr K M Levy	Medical Research Council
Dr J N Wingfield	Agricultural and Food Research Council

ANNEX 2.3

SCIENTIFIC VETERINARY COMMITTEE

The Scientific Veterinary Committee (SVC) was created by Commission Decision 81/651/EEC of 30 July 1981 to provide the EC Commission with expert advice on all scientific and technical problems concerning animal health, veterinary public health, and animal welfare. The membership of the sections concerning animal health and veterinary public health is described below:—

Animal Health Section

Chairman

Dr M K Eskildsen	Director, State Veterinary Institute for Virus Research, Lindholm
Dr R Ahl	Federal Research Institute for Virus Diseases of Animals, Tubingen
Professor J Badiola	Veterinary Faculty, Zaragoza
Dr G Dannacher	Deputy Director of the Laboratory of Bovine Pathology, Lyon
Dr F Garrido Abellan	Director, Laboratory of Animal Health of Granada, Santa Fe
Professor G Giorgetti	Institute of Animal Health, Basaldella
Professor F Guarda	Department of Animal Pathology, Faculty of Veterinary Medicine, University of Turin
Dr J Leunen	Director of the National Institute of Veterinary Research (retired), Brussels
Professor B Liess	Director of the Institute of Virology of the Veterinary College, Hanover
Dr J B McFerran	Veterinary Research Laboratories, Stormont, Belfast
Dr J P de Matos Aguas	Director of the National Veterinary Research Laboratory, Lisbon
Dr C Meurier	Director of Research, General Inspector of the Laboratories of the Veterinary Services, Central Veterinary Research Laboratory, Maison-Alfort
Professor G Radaelli	Faculty of Veterinary Medicine, School of the Public Veterinary Health Specialisation, University of Milan
Mr D J O'Reilly	Deputy Director, Veterinary Research Laboratory, Dublin
Professor O Papadopoulos	Professor of Microbiology and Parasitology, Veterinary Faculty, Thessalonika
Dr J Terpstra	Central Veterinary Institute, Lelystad
Professor J G Van Bekkum	Director of the Central Veterinary Laboratory (retired), Lelystad

Veterinary Public Health Section

Chairman

Professor D Grossklaus President of the Institute of Veterinary Medicine, Federal Office of Health, Berlin

Professor M Debackere Dean of the Veterinary Faculty, Director of the Institute of Pharmacology and Toxicology of Domestic Animals, Rijksuniversiteit, Gent

Dr G del Real Head of the Parasitology Service, Carlos III Health Institute, National Centre of Microbiology, Majadahonda (Madrid)

Professor K Gerigk Director of the Institute for Veterinary Medicine, Federal Office of Health, Berlin

Dr R J Gilbert Deputy Director, Central Public Health Laboratory, London

Dr F Janin Central Laboratory of Food Hygiene, Paris

Dr F Kenny Veterinary Inspector, Veterinary Research Laboratory, Dublin

Professor Ch.Labie Professor of Hygiene and of Foodstuffs of Animal Origin, National Veterinary School of Toulouse

Professor G E Lamming Professor of Animal Physiology, Department of Physiology and Environmental Studies, University of Nottingham, School of Agriculture, Loughborough

Professor P S Marcato Director of the Institute of General Pathology and Pathological Anatomy, Faculty of Veterinary Medicine, University of Bologna

Professor A Mandis Professor of Food Hygiene, Veterinary Faculty, Thessalonika

Professor A Osterhaus Department of Control of Virus Vaccines, Bilthoven

Professor A Penha-Goncalves National Veterinary Research Laboratory, Lisbon

Professor C Ring Institute for Food Science, Meat Hygiene and Technology of the Veterinary College, Hanover

Professor N P Skovgaard Royal Veterinary and Agricultural University, Copenhagen

Professor G Tiecco Institute of Food Inspection, Faculty of Veterinary Medicine, Bari

Dr D J Ventanas Barroso Veterinary Faculty, University Campus, Cáceres

ANNEX 2.4

RENDERING IN THE UNITED KINGDOM

THE RENDERING INDUSTRY

2.4.1 The rendering industry provides a service by collecting and processing all animal material from slaughterhouses, cutting plants and butchers' shops, which is not used for human consumption. Each year in the UK about 1.75 million tonnes of raw waste is removed and processed by renderers. About 300,000 tonnes of this is poultry material, mostly processed in dedicated poultry plants. The species of origin of the remaining material is estimated to be 54% cattle, 28% sheep and 18% pig. The products of the rendering process are fats and various protein meals. The annual output of the rendering industry is approximately 400,000 tonnes of protein meals, mainly meat and bonemeal and 250,000 tonnes of tallows.

THE RENDERING PROCESS

2.4.2 The rendering process involves crushing the raw material followed by the indirect application of heat. This evaporates the moisture and enables the tallow to be separated from the solids (greaves). The solids are then pressed, centrifuged or solvent extracted to remove the bulk of the remaining tallow before being ground into protein meals.

2.4.3 Depending on the raw material processed, tallows may be used for further refining for use in human food, soap manufacture, industrial purposes and animal feed. Protein meals (principally meat and bonemeal) are used in animal feedstuffs and a significant amount in pet foods. Inclusion rates for pig and poultry feed are between 2-5%.*

Types of rendering plants

2.4.4 Rendering plants can be divided into four main plants groups, depending on the raw material they process.

Technical Rendering

(a) The principle ingredient in technical rendering is low grade (green) offal and condemned material obtained from slaughterhouses — and other low grade material, which may contain fallen stock (or parts of) from knackers and hunt kennels. The tallow is used for industrial purposes, including lubricants, and some in animal feed. The meat and bonemeal produced by technical renderers is used in animal feed rations and petfood.

Rendering to produce high grade tallows

(b) The raw material is the fresh fat and bones obtained from slaughterhouses, cutting plants and butchers' shops. The tallow may be used for high quality toilet soap, for further refining, bleaching and deodorising for use in food manufacture, catering, etc. Animal protein is used in animal feed. Those processing plants producing tallow for human consumption are covered by the recently agreed Meat Products Directive and therefore detailed standards are laid down by Community law.

---

*The feeding of ruminant based animal protein to ruminants has been banned since 18 July 1988 in Great Britain and January 1989 in Northern Ireland. The feeding of animal protein derived from specified bovine offal (which is processed separately from other material) to any animal or bird, has been banned since 25 September 1990.

Edible Rendering

(c) There are a number of plants which utilise fresh kidney suet and channel (opening) fats for direct human consumption. The raw materials are usually processed at lower temperatures to produce beef dripping for frying and "premier jus" for packet suet and food manufacture. The greaves may be used in the manufacture of pet foods.

Specialist Plants

(d) There are a small number of plants which specialise in processing blood, feathers and poultry material. The products are used in animal feed and pet food.

Types of processing systems used

Batch Rendering

2.4.5 The traditional method of cooking is used by plants processing about 20% of the raw material. The average particle size of material entering the cookers is 40mm and average cooking time about $3\frac{1}{2}$ hours. Maximum temperatures range from 120-135°C under atmospheric pressure with a normal maximum of about 125°C. However one plant cooks under pressure (2 bar) for 35 minutes at 141°C. In some plants the load is discharged once the maximum temperature is reached, in others there may be a holding time of up to 20 minutes. On discharge the free run fat is drained off and the residual greaves removed for pressing and/or centrifugation to extract more fat. The dried greaves is subsequently ground to produce meat and bonemeal or sold as greaves to other renderers for further processing.

2.4.6 There are four continuous systems employed in the UK. They were first introduced in the early 1970s and now process about 80% of the raw material.

Stork Duke

The system works on the principal of a deep fat fryer. The rendering vessel operates with a high proportion of liquid fat, the heat being applied via a steam jacket and a steam heated tube rotor. The particle size of raw material entering the cooker is 20-50mm. Maximum temperatures achieved are between 135°C and 145°C with an average residence time of at least 30 minutes. The protein material is then pressed before being ground into meat and bonemeal.

Stord Bartz

There are a number of different types and sizes of Stord Bartz systems. The material (reduced to a particle size of 20-50mm) is heated by a steam heated disc rotor. The discs occupy the length of the rendering vessel. The average maximum temperature achieved is around 125°C and an average residence time of between 22 and 35 minutes. Pressing and grinding is similar to the procedure operated in the previously described system.

Anderson Carver-Greenfield

Finely minced (to a size of less than 10mm) raw material is first mixed with recycled heated tallow to form a slurry. This is then pumped through a system of tubular heat exchangers with vapour chambers under partial vacuum before being centrifuged and pressed. The heat treatment involves a maximum process temperature of 125°C with an average residence time of between 20 and 25 minutes.

Raw material is initially minced to a particle size of 10 mm before being heated to 95°C for 3-7 minutes. The liquid phases (fat and water) are removed by centrifuging or light pressing and are further separated to recover the tallow. The resultant solids are dried at temperatures ranging from 120°C to 130°C.

DIRECTIVE 90/667 EEC ON THE DISPOSAL OF ANIMAL WASTE

2.4.7 Under this Directive, which comes into force in 1992, all rendering plants have to meet structural and other requirements and must comply with specific microbiological standards, including for salmonella. Community trade in the products of rendering will only be possible from approved plants. The likely impact on the UK industry is further rationalisation.

ANNEX 2.5

FISHMEAL MANUFACTURING IN THE UNITED KINGDOM

FISHMEAL USAGE

2.5.1. In the UK each year, approximately 300,000 tonnes of fishmeal, 85% of which is imported from a range of countries including Chile, Iceland, Norway, Denmark and Peru and subject to the requirements of the Importation of Processed Animal Protein Order 1981, is used in the manufacture of animal feedingstuffs. Of this total about 60% is used in poultry rations, 10% in pig rations, 5-10% in feed for ruminants, 10% for aquaculture, and the remainder for other uses.

UK FISHMEAL INDUSTRY

2.5.2. The UK fishmeal industry manufactures about 45,000 tonnes of fishmeal a year and is based on four plants in Aberdeen, Grimsby, Shetland and Fleetwood, all of which are registered under the Processed Animal Protein Order 1989. UK production represents about 0.8% of total world production of fishmeal.

The fishmeal production process

2.5.3. In the production processes used in the UK, raw material (ie fish offal from fish plants and fish surplus to human consumption) is heated to 80-100°C in a cooker before going through a press which divides the cooked material into press cake and press water. Press water is passed to a decanter where solids are separated. The solids or "grax", 33% of which is dry matter, are mixed with the press cake for final drying in the drier. The press water from the decanter passes through a separator where free oil is separated from the remaining press water to produce fish oil. The remaining water (containing 10% protein) from the separator is then concentrated into solubles by evaporation using waste heat from the processing system. The solubles are then subjected to final drying together with the "press cake" and "grax" which brings the moisture content to below 10%. The meal is then stable and goes for bagging or bulk storage.

2.5.4. $4\frac{1}{2}$ tonnes of raw material yields 1 tonne of fishmeal and $\frac{1}{2}$ tonne of oil. The best quality oil is used in feed for salmon; the rest is used in human food.

DIRECTIVE 90/667 EEC ON THE DISPOSAL OF ANIMAL WASTE

2.5.5. Under the Directive fish offal is classed as "low risk" and all fish meal plants will have to meet specified structural and operational requirements and the final product must meet prescribed microbiological standards, including those for salmonella. Intra-Community trade in fishmeal will only be possible for fishmeal which comes from plants approved in accordance with the Directive. Similar microbiological standards will apply to fishmeal imported from third countries.

CHAPTER 3

SALMONELLA AND OTHER PATHOGENS

INTRODUCTION

3.1 This chapter is concerned predominantly with salmonella and, within this group of organisms with a particular strain, *Salmonella enteritidis* phage type 4, which has been responsible for the recent serious rise in salmonella infections in man in the UK. Although we have focused much of our attention on this organism we have considered salmonella infections in general. Our recommendations provide a basis for the prevention and control of possible future problems due to other salmonella strains.

3.2 Other organisms considered include anthrax, listeria, yersinia, *E.coli* 0157 and campylobacter.

3.3 The microbiological safety of food has come under increasing scrutiny in recent years. The Report of the Committee on the Microbiological Safety of Food (The Richmond Committee) was published in 1990. The report recommended setting up an Advisory Committee on the Microbiological Safety of Food to provide an independent expert view on the public health implications of food safety matters and in particular on the results of surveillance. Further it recommended setting up a Steering Group on the Microbiological Safety of Food to coordinate microbiological food safety. The Committees were formed in 1990. The membership and terms of reference of the Steering Group is shown at Annex 3.5. We are aware of the work of these and other Committees in this area and we have been informed of the projects they are undertaking. This was helpful for our deliberations.

3.4 We regard animal feedingstuffs as an integral part of the feed/farm animal/human food chain for the purposes of ensuring food safety and consider that the existing arrangements for ensuring the microbiological safety of human food are such that animal feed could at least in part, be incorporated into the existing arrangements. Some of our recommendations stem from this approach.

SALMONELLA INFECTIONS OF MAN

3.5 Salmonella has been an important cause of human food poisoning for many years resulting in considerable costs and loss of productivity. The early 1980s saw the start of the rise of *S. enteritidis* infections which has subsequently become very important. The data, illustrated in Annex 3.1, shows that between 1981 and 1991 the annual number of *S. enteritidis* isolations from humans rose from 392 to 14,693.

ROUTES OF TRANSMISSION OF SALMONELLA

3.6 Routes of transmission of salmonella in animals and in man are discussed in the Report of the Committee on the Microbiological Safety of Food (Part I) (Appendix I Salmonellosis). The Report states "Transmission to man is usually foodborne from infected food animals. . . ." Because there is evidence linking poultry meat and eggs with the rise in human *S. enteritidis* infections we have concentrated much of our attention on the poultry industry and on *S. enteritidis*.

SALMONELLA IN POULTRY MEAT AND EGGS

3.7 Work undertaken by the Public Health Laboratory Service (PHLS) in 1990 showed that 48% of home-produced chilled and frozen chickens were contaminated with salmonella, and in 21% the contamination was *S. enteritidis*. Salmonella contamination of imported chicken was found to be 72% in a survey carried out in 1991, and 16% of chickens contained *S. enteritidis*. Imported chickens form approximately 12% of the market. A small survey carried out by PHLS found 0.6% of eggs from salmonella infected flocks to be internally contaminated. The contamination rate at retail level will be lower because there will be dilution with eggs from uninfected flocks. A number of surveys of both home produced and imported eggs are in progress and will be considered by the Advisory Committee on the Microbiological Safety of Food and its Working Group on Salmonella in Eggs.

SALMONELLA INFECTIONS OF POULTRY

Poultry production

3.8 The structure of, and operations within, the poultry industry may have important implications for the way in which *S. enteritidis* has established itself.

3.9 The poultry industry may be described (see Annex 3.3) as consisting of two pyramids each with elite flocks at the apex of the pyramid and either layer flocks or broiler flocks at the base. Elite broiler flocks are maintained in the UK but elite layer flocks are maintained mainly in France and Holland. Because *S. enteritidis* is able to contaminate the contents of the intact egg prior to laying, infection may be spread vertically within poultry flocks so that any infection occurring at elite, grandparent or parent level will be important.

Broiler breeders

3.10 There are currently 622 broiler breeder premises registered with the Ministry of Agriculture Fisheries and Food (MAFF). Since controls were introduced *S. enteritidis* has been confirmed in flocks on 108 premises but some may be repeat isolations from the same site, 58 flocks have been slaughtered because of evidence of vertical transmission.

Layer breeders

3.11 Of the 172 layer breeder premises registered with MAFF *S. enteritidis* has been confirmed on 11 occasions at 8 premises. In 1991 35,756 layer breeders were slaughtered.

Broilers and layers

3.12 In 1991 *S. enteritidis* was confirmed in adult laying birds on 61 holdings (48 with flocks of 25+) and 179,828 layers were killed. Broiler birds are not subject to statutory monitoring.

SOURCES OF INFECTION IN POULTRY FLOCKS

3.13 The three major potential sources of infection in poultry flocks are parent birds, the environment and feedingstuffs. Routes of infection are shown at Annex 3.2. Clearly the breeding bird can be a source of initial infection although its relative importance as a source is unclear. In addition, the environment may become heavily contaminated and may be a cause of further transmission. However, it is difficult in the current *S. enteritidis* outbreak to assess the contribution of feedingstuffs relative to other sources at the present time. This appears to be a well established zoonosis in which an infection, possibly introduced by feedingstuffs, has resulted in further vertical transmission and widespread environmental contamination.

SALMONELLA IN OTHER ANIMALS

Cattle

3.14 Incidents in livestock are reportable under the Zoonoses Order. The predominant serotype isolated from all cattle is *S. typhimurium* followed by *S. dublin*. In calves *S. typhimurium* is the commonest serotype and the most prevalent phage type in 1990 was PT 204c. (See also note on increased multiple antibiotic resistance *S. typhimurium* paragraph 4.26). *S. enteritidis* is sometimes isolated from cattle but field investigations often implicate poultry and in particular poultry litter as the source of infection. Human infections with *S. typhimurium* have varied from around 4,000 in 1981 to almost 8,000 in 1983 and 1987 and stabilised around 5,000 in 1990 and 1991.

Sheep

3.15 Salmonella isolations from sheep are uncommon compared with cattle. Serotypes most frequently isolated are *S. arizona*, *S. montevideo* and *S. typhimurium*. There is little information linking human infection of salmonella with sheep.

Pigs

3.16 Serotypes most frequently isolated from pigs are *S. typhimurium*, *S. derby* and *S. kedougou*. There has been a steady rise of *S. typhimurium* phage type 193. In 1990 this type represented almost 11% of all salmonella isolates from pigs and was the commonest *S. typhimurium* phage type isolated from humans. However this phage type is also isolated from cattle.

SALMONELLA IN FEEDINGSTUFFS AND RAW MATERIALS

3.17 Data on salmonella in feedingstuffs is available from a number of sources; mandatory tests of animal protein carried out by MAFF and by the processors themselves, and reports by laboratories of positive results obtained from samples submitted voluntarily by suppliers and manufacturers in the feed chain. Some feedingstuffs surveys have also been carried out. Information from these various sources is given in paragraphs 3.18 to 3.24.

Domestically produced animal protein

3.18 Official testing of animal protein is carried out by MAFF. In 1989 1,674 samples were taken, 5% were positive for salmonella and eight samples for *S. enteritidis*. In 1990 comparable figures were 1,388, 3% and one respectively. In 1991 provisional figures are 1,360, 3% and two.

Imported supplies of animal protein

3.19 The incidence of salmonella contamination of imported animal protein is generally higher than that for home-produced material. Frequency of testing varies with increased testing when contamination persists in particular materials. In 1990 of 2,011 samples of imported material tested by MAFF 16.8% was contaminated with salmonella, and five samples contained *S. enteritidis*. Provisional figures for 1991 for 2,089 samples show salmonella isolation rates of 12.5% and with no *S. enteritidis* isolated.

Imported fish protein

3.20 We noted that levels of contamination were now generally low although high isolation rates have sometimes been reported in material from S. America. A study by the Norwegian authorities following the detection, at point of entry, of contamination of Norwegian fish meal suggested that contamination at the port may occur.

Vegetable raw materials

3.21 Home-produced and imported supplies of vegetable materials are not subject to the same statutory sampling and testing regimen as animal protein. They are however subject to the voluntary Codes of Practice for the Control of Salmonella in Raw Materials. Analysis of mandatory reports under the Zoonoses Order shows that the main contaminated vegetable materials were oilseed proteins — cotton, rape, palm kernel, soya and sunflower. The feedingstuffs industry was concerned about the level of contamination particularly from home-produced rape.

3.22 A MAFF survey of salmonella contamination of raw materials carried out in the first four months of 1990 confirmed the pattern of reports under the Zoonoses Order referred to in paragraph 3.21. The survey undertaken at compounders premises showed that of 1,465 samples 7.1% were contaminated. Where comparisons could be made imported materials were generally more contaminated than home-produced. The commonest serotype isolated was *S. senftenberg*. *S. typhimurium* was isolated on 12 occasions, and *S. enteritidis* was not isolated. A second survey, involving 1,144 samples carried out by industry at vessel discharge showed an overall incidence of contamination of 2.8%. The difference in level of isolations recorded by the trade on vessel discharge (2.8%) and by MAFF in store (7.1%) raises the possibility that contamination in transport or storage may be a problem.

Straight feedingstuffs

3.23 Because straight feedingstuffs (as described in Chapter 5) are not subject to statutory controls or codes of practice for salmonella there is no information about the incidence of salmonella contamination. **We can see no reason for excluding straight feedingstuffs from the controls which apply to raw materials. We recommend that straight feedingstuffs should be subject to a code of practice.**

Finished feedingstuffs

3.24 The reports from laboratories under the Zoonoses Order of positive samples from feedingstuffs and feed ingredients (including vegetable and animal proteins) show that in 1990 the total number of salmonella isolations was 1,325 of which ten were *S. enteritidis*. In 1991 salmonella were isolated in 1,971 occasions of which none were *S. enteritidis*.

SALMONELLA CONTROL MEASURES Control measures in animals

3.25 Because salmonella are widespread in animals and their surroundings the controls introduced in 1989 were of necessity wide ranging. They were designed to introduce continuous bacteriological monitoring, to remove infected stock, to ensure replacement poultry stock was not infected and arrived at clean premises, and to prevent the introduction of infection through feedingstuffs. The controls consist of legislative and voluntary measures. Annex 3.4 gives details of the regulations and the codes of practice.

Poultry flock owners

3.26 Flocks of 25 or more breeding birds, and of 100 or more laying birds have to be registered with MAFF; flocks of 25 or more laying or breeding birds, and flocks of less than 25 birds if their eggs are sold for human consumption, have to be tested for salmonella, and to observe certain hygiene standards. Voluntary codes of practice complement these statutory obligations by setting out requirements for post-slaughter cleaning and disinfection and the need to show that premises are free from contamination before restocking.

Requirements for animal protein processors and importers

3.27 Processors of animal protein must be registered, the processed products must be sampled, and tested. Imports of animal protein, including fishmeal, must be licensed by MAFF. The licence conditions are adapted in the light of the results of bacteriological monitoring. Positive results must be submitted to MAFF.

Requirements for raw material suppliers and feedingstuffs manufacturers

3.28 Suppliers of animal protein and vegetable materials, and feedingstuffs manufacturers, including on-farm mixers have agreed to comply with voluntary codes of practice which give guidance on the selection, handling and storage of raw materials to minimise contamination.

ENFORCEMENT

3.29 The regulations which apply to animal protein are enforced jointly by local authorities and MAFF officials. MAFF licences consignments of imported material and takes samples for salmonella testing. Animal protein processors in Great Britain are registered with MAFF and their products are sampled and tested by MAFF as a check on the sampling and testing regimen carried out by the processor. Local authority inspectors check that premises are properly registered and take action if there are contraventions. There appear to be no great difficulties with this arrangement.

REPORTS UNDER THE ZOONOSES ORDER

3.30 The effectiveness of the statutory controls and the voluntary codes on raw materials and feedingstuffs can to an extent be measured by reviewing data on the incidence of salmonella in the feed/food chain. We have described the results of isolations in humans, animals and animal protein in earlier paragraphs. We were also informed that the majority of isolations reported under the Zoonoses Order are from samples provided by large commercial compounders and integrated units. Of the isolations reported in 1990 and in the first four months of 1991 50% were from the same five feed compounders. The remainder came from smaller commercial compounders. There is no simple way of determining how many isolations may have come from on-farm mixers. This makes it impossible to say whether the reports under the Zoonoses Order are in any way representative of the total feed supply. **We are particularly concerned about the absence of information on farm-mixed feed. We recommend that a survey be initiated to determine the microbiological quality of feed produced on farms.**

3.31 Reports under the Zoonoses Order relate only to positive samples, reported by the laboratory which identified them. The figures cannot be measured against a denominator figure for the number of samples submitted for analysis nor can all the reports be traced to source. We have in our discussions agreed that it is important to have figures for negative results. **We recommend that laboratories report positive and negative salmonella results for feedingstuffs, and the source of the feedingstuffs, to MAFF and that these results be evaluated by the Advisory Committee on Animal Feedingstuffs.**

REPORTS UNDER THE PROCESSED ANIMAL PROTEIN ORDER

3.32 We note (in paragraph 3.17) that 5% of samples of domestically-produced animal protein were positive for salmonella in 1989 and 3% in 1990 and 1991. Comparable figures for imported animal protein were 16.8% in 1990 and 12.5% in 1991. **We support the continuing efforts being made to reduce contamination and we recommend that the current controls be retained.**

FEED SURVEYS AND RESEARCH

3.33 It is difficult to draw conclusions on the relative importance of raw materials and feedingstuffs as sources of contamination since the individual surveys (referred to in paragraph 3.22) have not been conducted in a similar manner and may not have used the same techniques. This in our view strengthens the need for ongoing, comparable, survey work. There are a number of recommendations in the Report by the Committee on the Microbiological Safety of Food which advise a continuing critical scrutiny of the quality of processed animal protein, vegetable protein and finished feedingstuffs (Recommendations 6.6, 6.7, 6.8, 6.9). The report envisaged the involvement of the Steering Group on the Microbiological Safety of Food in these investigations. **We recommend that further surveys be undertaken to monitor feedingstuffs and vegetable materials and that the results be made available to the Animal Feedingstuffs Advisory Committee.**

3.34 We have been told that MAFF is carrying out research work to identify levels of salmonella contamination in small compounder's premises, at milling and crushing plants and on some farms. When completed it should provide valuable information about the situation in smaller units. The results will provide MAFF with information which is not readily identifiable from the Zoonoses Order. The survey will enable an overview to be taken of all types of animal feed. **We recommend that the results of these investigations at compounders, crushing plants and farms should be examined by the Animal Feedingstuffs Advisory Committee.**

FEED AS A CONTRIBUTOR TO SALMONELLA CONTAMINATION IN FOOD

3.35 There are serious difficulties in trying to assess the contribution feed may make to the carriage of salmonella by animals. The predominant salmonella serotype found in feedingstuffs and feed ingredients is not *S. enteritidis*, which is the one most often found in poultry meat and eggs, and associated with human infection. There are however examples, from previous outbreaks, of salmonella serotypes introduced in feedingstuffs causing animal and human infection. *S. agona* was isolated from Peruvian fish meal in 1970 and soon after isolates were obtained from poultry and humans. Pigs also became infected. It was considered that a cycle of infection independent of the original fish meal had been set up. Salmonella 4.12:d: — (monophasic) was isolated from fish meal imported from South Africa; the strain was also found in poultry offal meal and feather meal, and both pigs and poultry became infected. Human infections occurred in 1968 and 1969 with a declining number in 1970 and 1971. Although there are examples such as these there is generally insufficiently robust data on salmonella contamination in feedingstuffs to allow a link to be established between previous feed contamination and current infection/contamination in animals. Further surveys using consistent methods and techniques and fuller information accompanying reports under the Zoonoses Order are needed to provide information on the incidence of carriage and disease in animals, and also on the relationship of this to the incidence of salmonella in feedingstuffs.

IMPORTANCE OF UNCONTAMINATED FEED

3.36 The consensus view presented to us is that feed is important in introducing contamination into flocks and that every effort should be made to reduce contamination from this source. Not least is the argument that all efforts to disinfect houses and maintain clean flocks are negated if infection is reintroduced by feedingstuffs. This argument, sustained by all the experts we met is supported by the examples cited in 3.35. We have been told that the impact of contaminated feed will be greater if it enters breeding flocks and that concentrated efforts to reduce contamination at that point would have maximum effect. Nevertheless the removal of feed contamination alone at this stage of the epidemic would not necessarily have a significant effect on the current *S. enteritidis* disease in the human population.

FEED TREATMENTS

3.37 The effectiveness of various approaches to destroying salmonella in feed have been investigated.

Pelleting

3.38 Broilers are usually fed pelleted feed and layers fed meal. The pelleting process generally creates a temperature high enough to kill or reduce the numbers of salmonella. However, contamination downstream from the presses in the mill is not uncommon. We are aware that in some EC member states all poultry feed must be heat treated. Heat treatment (with prevention of post pasteurisation contamination) appears a potentially useful method of eliminating salmonella in feed. **We recommend that the feasibility of heat treating all poultry feed should be examined by the Animal Feedingstuffs Advisory Committee. Priority should be given to the possibility of heat treating feed supplies destined for elite and grandparent flocks.**

Irradiation

3.39 Irradiation can be effective. There are no legal obstacles but cost is a considerable deterrent.

Acid treatments

3.40 There are some acids permitted for use in finished feedingstuffs. They are listed in EC Directive 70/524 and the Feeding Stuffs Regulations 1991 (discussed in Chapter 5). There are no restrictions on the use of acids in raw materials. We have been informed that the use of acids and their salts in the treatment of feed is receiving more attention. However, a number of problems have been drawn to our notice including unpalatability, corrosion of equipment and potential harm to operators. The importance of these problems is not well defined. **We recommend that the Animal Feedingstuffs Advisory Committee should examine this area in some detail with the objective of trying to define a system which could be applied throughout the industry.**

COLLECTION AND PUBLICATION OF INFORMATION ON THE INCIDENCE OF SALMONELLA

3.41 A number of publications report isolations of salmonella in humans, food, animals, and animal protein. The PHLS/SVS Salmonella Update which is issued quarterly summarises animal and human infections. A separate annual publication, the Animal Salmonellosis Summary, contains the reports of MAFF tests carried out under the statutory authority of the Processed Animal Protein Order and the Imported Processed Animal Protein Order and also positive isolations in animals which are notified under the Zoonoses Order. Results in raw materials and animal feedingstuffs notified under the Zoonoses Order are not published. There is no single publication which assembles all the data on infection/contamination in humans and in the feed/food chain and some data is not published at all. **We have already recommended fuller reporting under the Zoonoses Order. We recognise that there are difficulties in producing strictly comparable data between different parts of the feed, animal, human food and human infection chain. However we recommend that every effort be made to provide data which is obtained using methods which are as similar as possible. This will be particularly valuable if national surveillance results have to be reported to the EC Commission as part of the harmonisation of salmonella controls. We recommend that in addition to the existing publications consideration be given to publishing all data on salmonella infection/ contamination in a single publication.**

DEVELOPMENT OF POLICY ON SALMONELLA

3.42 We have emphasised in this Chapter that more and better coordinated information on salmonella in raw materials and feedingstuffs will provide a basis for further assessment of the significance of feed contamination in relation to infection in animals and humans. This will also provide a sound basis for considering what measures may need to be taken to reduce feed contamination. These issues can appropriately be examined by the Animal Feedingstuffs Advisory Committee whose proposed functions are described in Chapter 7. **We recommend that the Animal Feedingstuffs Advisory Committee should consider all available information on salmonella and other pathogens in feedingstuffs with particular reference to the possibility that serotypes found in feed may subsequently become important for animals and man, and advise on any measures which need to**

be taken. Further we recommend that authorities charged with monitoring salmonella and other organisms in animal feed, animals, food, and humans be represented on this Committee.

OTHER PATHOGENS
Anthrax

3.43 Anthrax is primarily a disease of herbivores. It caused heavy losses in cattle, sheep and goats throughout the world until an effective veterinary vaccine was developed in the 1930s. The incidence in the UK and in many other developed countries has been reduced to very low levels.

3.44 Humans normally contract the disease through contact with herbivores with the disease, or from products such as hides, hair, wool, and bone meal derived from affected animals. The most common manifestation of the disease is cutaneous anthrax, although intestinal and pulmonary forms also occur. Reported cases of human and animal anthrax in Great Britain 1981-1991 are shown at Annex 3.6.

3.45 In 1977-1978 a major series of outbreaks occurred in cattle in which imported groundnut was implicated. Feed was suspected in a major outbreak in pigs in 1989 but *B. anthracis* was not found in the feed samples examined.

3.46 The risks from anthrax appear to be less now than formerly. However there is need for a continuing vigilance, particularly as regards feed ingredients from countries where the disease is known to be endemic. We are satisfied with the present legislation.

Listeria

3.47 The bacterium *Listeria monocytogenes* has been recognised for many years as a human and animal pathogen. It is widespread in the environment being present in sewage, slurry, soil, surface water, vegetation and silage. It can be transmitted via food and has been isolated from a range of foods.

3.48 Listeriosis occurs in animals and is of greatest economic importance in sheep and cattle but it may also occur in other species such as goats, pigs, horses and birds. Figures obtained from diagnoses recorded at Veterinary Investigation Centres in England, Scotland and Wales show that listeriosis in sheep increased between 1979 and 1989 but reports from other animals show little change. The increase in ovine listeriosis is thought to be partly due to increased use of silage as a feedingstuff, including greater use of "big-bale" silage and partly to increased "in-wintering". There is little information about listeria in animal feedingstuffs other than silage.

3.49 In adult humans listeriosis has a range of clinical presentations ranging from a mild 'flu-like illness to a severe disease. The number of reported cases in England and Wales increased between 1967 to 1988 and there was a sharp increase in 1987 and 1988 with a down turn in the latter part of 1989 (see Annex 3.7). Following preventative action the reported incidence of human listeriosis in England and Wales is now at the low levels seen in the early 1980s.

3.50 There is no clear evidence of a direct link between animal and human listeriosis apart from a few individual localised skin infections among veterinary surgeons and farmers caused by contact with infected animals. Current work on subtyping of isolates which we support should demonstrate whether there is commonality between feed, animal, and human strains.

OTHER INFECTIONS

Campylobacter
Yersinia
E. coli 0157

3.51 In the case of listeria and anthrax there is some, albeit limited, information about the role of feedingstuffs. However, for the three remaining organisms we have considered, campylobacter, yersinia and *E.coli* 0157, we have not found any information about their occurrence in feedingstuffs. Indeed, the relationship of human disease to animal carriage is less clear than is the case with salmonella. **Campylobacters are a major cause of human** intestinal disease (see Annex 3.8), *E.coli* 0157 infection has a significant mortality and important long term morbidity in children, and yersinia is of increasing importance. For these reasons we recommend that following further work on the relationship between animal carriage and human disease, the role of animal feedingstuffs should, as appropriate, be considered.

ANNEX 3.1

CURRENT ISOLATIONS IN HUMANS, ANIMALS, POULTRY AND PROCESSED ANIMAL PROTEIN

Table 1: Salmonella in Humans – England and Wales

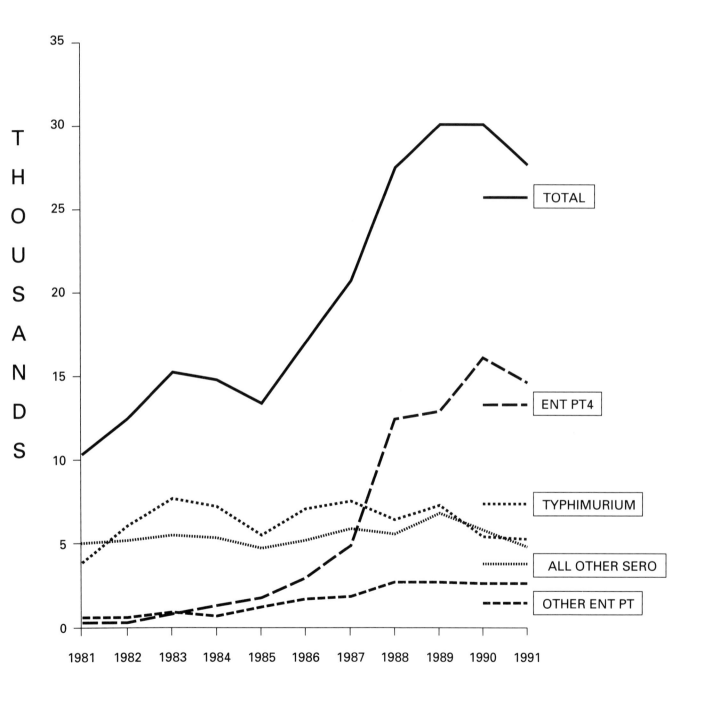

Source: Division of Enteric Pathogens
Public Health Laboratory Service

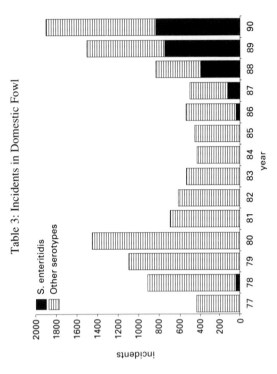

Table 2: Incidents in Food Animals

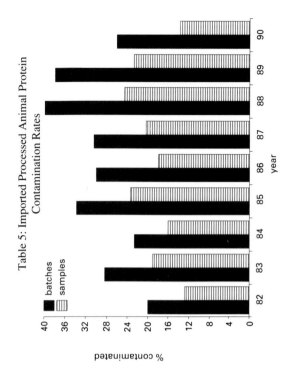

Table 3: Incidents in Domestic Fowl

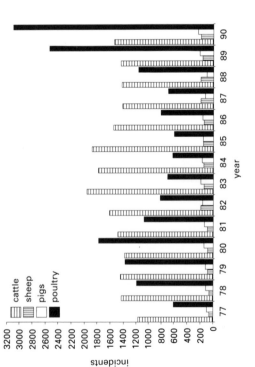

Table 4: Domestic Processed Animal Protein
Contamination Rates

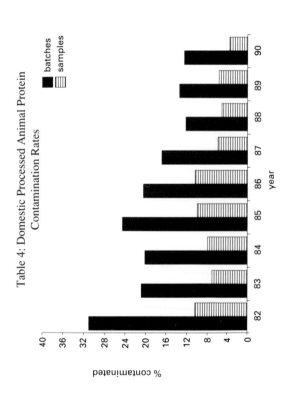

Table 5: Imported Processed Animal Protein
Contamination Rates

Note: An incident is defined as the reported identification of salmonella from an animal or group of animals, an animal carcase, or from animal products or surroundings which can be specifically related to identifiable animals and from animal feeds. If more than one animal species is identified from a common source, separate incidents are reported for each species.

Source: MAFF, ADAS: Animal Salmonellosis 1990

32

ANNEX 3.2

ROUTES OF INFECTION FOR SALMONELLA

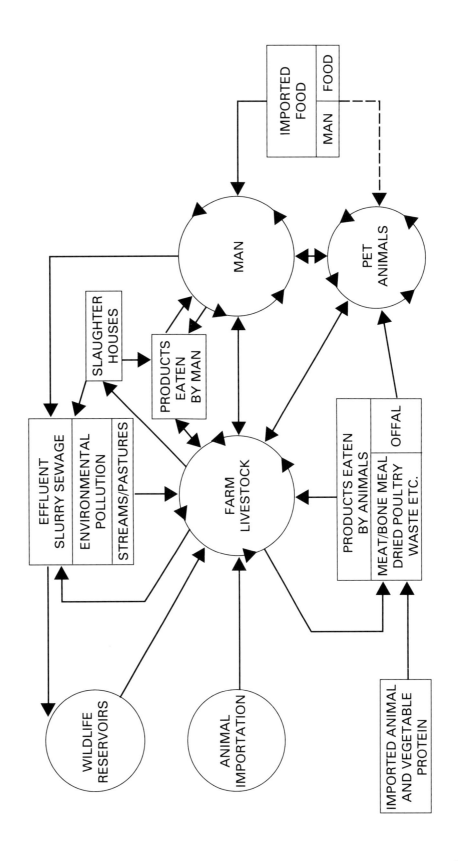

STRUCTURE OF THE UK POULTRY AND EGG INDUSTRY

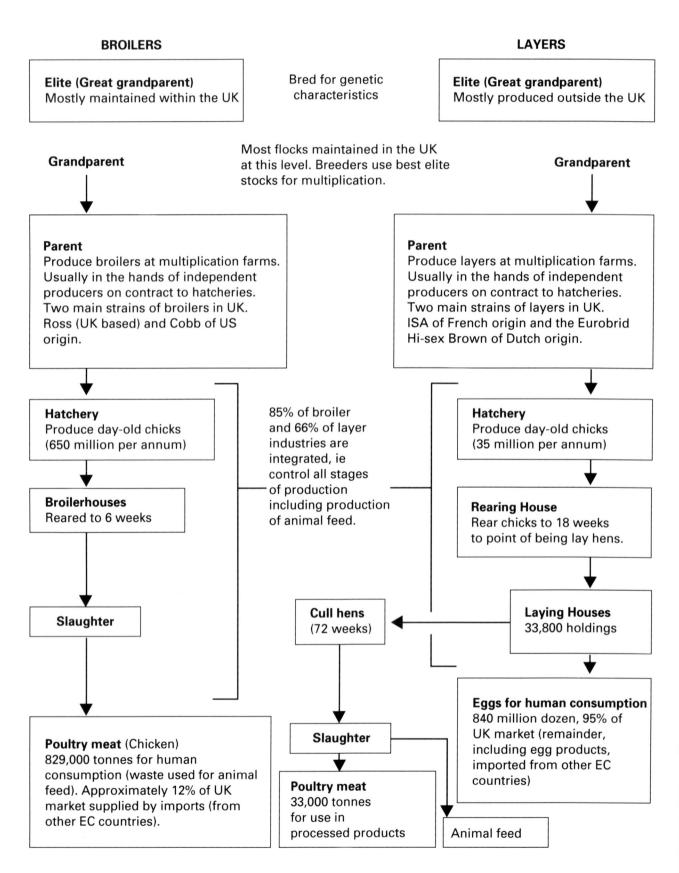

BROILERS

Elite (Great grandparent)
Mostly maintained within the UK

Bred for genetic characteristics

LAYERS

Elite (Great grandparent)
Mostly produced outside the UK

Grandparent

Most flocks maintained in the UK at this level. Breeders use best elite stocks for multiplication.

Grandparent

Parent
Produce broilers at multiplication farms. Usually in the hands of independent producers on contract to hatcheries. Two main strains of broilers in UK. Ross (UK based) and Cobb of US origin.

Parent
Produce layers at multiplication farms. Usually in the hands of independent producers on contract to hatcheries. Two main strains of layers in UK. ISA of French origin and the Eurobrid Hi-sex Brown of Dutch origin.

Hatchery
Produce day-old chicks
(650 million per annum)

85% of broiler and 66% of layer industries are integrated, ie control all stages of production including production of animal feed.

Hatchery
Produce day-old chicks
(35 million per annum)

Broilerhouses
Reared to 6 weeks

Rearing House
Rear chicks to 18 weeks to point of being lay hens.

Slaughter

Laying Houses
33,800 holdings

Cull hens
(72 weeks)

Poultry meat (Chicken)
829,000 tonnes for human consumption (waste used for animal feed). Approximately 12% of UK market supplied by imports (from other EC countries).

Slaughter

Eggs for human consumption
840 million dozen, 95% of UK market (remainder, including egg products, imported from other EC countries)

Poultry meat
33,000 tonnes for use in processed products

Animal feed

ANNEX 3.4

LEGISLATION: SALMONELLA AND OTHER PATHOGENS

PRIMARY LEGISLATION

3.4.1 The Animal Health Act 1981 in Great Britain, and the Diseases of Animals (Northern Ireland) Order 1981 provide the enabling powers for the statutory controls which are designed to ensure the microbiological safety of feedingstuffs. The relevant regulations enacted under these powers are listed and summarised in the following paragraphs. The parallel legislation for Northern Ireland is listed in Para 1.5 (below).

THE ZOONOSES ORDER 1989 (SI 285)

3.4.2 This order came into force in Great Britain on 1st March 1989 and revoked the Zoonoses Order 1975. The Order extended reporting under the 1975 Order (cattle, sheep, goats, pigs, rabbits, domestic fowls, turkeys, geese, ducks, guinea-fowl, pheasant, partridges and quails) by adding horses, deer and pigeons to the list. Under the 1989 Order the responsibility for reporting the identification of salmonella has been placed on the person carrying out the examination or the person in charge of the laboratory carrying out the examination. All isolations of salmonella from samples taken from an animal or bird, or from the carcase, products or surroundings of an animal or bird or from any feedingstuffs must be reported to a veterinary officer of the Ministry of Agriculture, Fisheries and Food (MAFF), normally the Senior Investigation Officer at one of the Ministry's Veterinary Investigation Centres in England and Wales, or a Divisional Veterinary Officer in Scotland. The Order allows veterinary inspectors to enter any premises and carry out such enquiries as are considered necessary to determine whether salmonella is present. Powers are also available to declare a premises infected, to prohibit the movement of animals, poultry, carcases, products and feedingstuffs in or out of the premises except under licence, and to serve notices requiring the cleansing and disinfection of premises where salmonella is known to have been present.

THE PROCESSED ANIMAL PROTEIN ORDER 1989 (SI 661)

3.4.3 This Order came into effect in Great Britain on 14th April 1989 and revoked The Diseases of Animals (Protein Processing) Order 1981. It provides for the statutory registration with MAFF of processors who subject animal protein to a process which converts it into a state in which it could be used (with or without further treatment or procedure) as a feed ingredient for livestock or poultry. It also provides for the testing of all processed animal protein for salmonella. Animal protein is any material which contains the whole or any part of any dead animal or bird, or of any fish, reptile, crustacean or other cold-blooded creature or any product derived from them and includes blood, hatchery waste, eggs, egg shells, hair, horns, hides, hoofs, feathers and manure, any material which contains human effluent, and "processed animal protein" any protein obtained from any of these materials by heat, sedimentation, precipitation, ensiling or any other system of treatment or procedure but does not include milk or milk products, shells other than egg shells, fat or dicalcium bone phosphate. Registered processors are required to take samples of processed animal protein on each day that it is dispatched from their premises, or is incorporated into feed on the premises, and have them tested for salmonella at laboratories authorised by MAFF using one of the methods detailed in Schedule 1 to the Order. This daily testing is carried out at the processor's expense. Agricultural Department officers also visit registered premises over 20 days each year to take samples of processed animal protein produced and test them for salmonella. On premises where material is produced by a continuous process, batches of samples, statistically related to the volume of production, are collected twenty days during the year. The Order also imposes a duty on registered persons who know that a test carried out on a sample of processed animal protein taken from their premises has isolated salmonella, to ensure that no further protein is moved from the same storage facility until it has been tested and found negative for salmonella or is moved under a licence authorised by the appropriate Agricultural Department.

THE IMPORTATION OF PROCESSED ANIMAL PROTEIN ORDER 1981 (SI 667) (AS AMENDED)

3.4.4 The Order prohibits the landing of processed animal protein into Great Britain except under a written licence issued by the appropriate Minister. More rigorous licensing conditions for imported animal and fish protein were introduced on 1 May 1989. Imported processed animal protein is also monitored for salmonella, the frequency of sampling takes account of the testing history of the type of material and the country of origin. The number of samples taken from each consignment is statistically related to the quantity of the consignment.

3.4.5 The above legislation is enforced primarily by local authorities but some enforcement responsibility also rests with the appropriate Agricultural Departments.

LEGISLATION ENACTED UNDER THE DISEASES OF ANIMALS (NORTHERN IRELAND) ORDER 1981

3.4.6 The secondary legislation in Northern Ireland largely reflects that of Great Britain and is contained in parallel orders — the Zoonoses Order (Northern Ireland) 1991 (SR 305), the Diseases of Animals (Animal Protein) (No 2) Order (Northern Ireland) 1989 (SR 347) and the Diseases of Animals (Importation of Processed Animal Protein) Order (Northern Ireland) 1989 (SR 286), respectively. Unlike in Great Britain local authorities have no role in the enforcement of the above legislation. Responsibility for enforcement lies with the Department of Agriculture for Northern Ireland (DANI).

CODES OF PRACTICE

3.4.7 These are non-statutory controls designed to control salmonella in feedingstuffs and feed raw materials. They were introduced in 1989 following consultation with the industry. One of the main requirements in all the codes is regular monitoring using the same methods as those for animal protein under the Processed Animal Protein Order (see above). The Codes of Practice in place in Great Britain are as follows:—

(a) *Code of practice for the control of salmonella during the storage, handling and transportation of raw materials intended for incorporation into animal feedingstuffs.* The purpose of this Code is to ensure that raw materials supplied for incorporation into animal feedingstuffs are of a satisfactory bacteriological standard with the risk of salmonella contamination minimised.

(b) *Code of practice for the control of salmonella in the animal by-products rendering industry.* This supports the legislation (the Processed Animal Protein Order) and requires that any protein meals produced from raw materials of animal or poultry origin and intended for incorporation into feedingstuffs for livestock or poultry are processed at premises registered with the Ministry, the method of processing is sufficient to destroy salmonella and that no contamination of the processed products takes place on the registered premises following processing.

(c) *Code of practice for the control of salmonella for the UK fishmeal industry.* This has similar provisions to those for the animal by-products industry.

(d) *Codes (2) of practice for the control of salmonella in the production of final feed for livestock in premises producing (i) over 10,000 tonnes per annum and (ii) less than 10,000 tonnes per annum.* The purpose of these Codes is to provide guidelines for establishing good practices for the production of finished feeds for livestock. These follow the production process through from beginning to end, specifying acceptable sources of raw materials, the hygiene training which personnel should have undertaken, the design and structure of manufacturing premises and conditions of transport for the final feed.

3.4.8 In Northern Ireland there are two codes of practice for the control of salmonella in feedingstuffs, during:—

(a) the storage and transport of raw materials and animal feedingstuffs, and

(b) the manufacture of animal feedingstuffs. Their purpose is the same as that of the Codes in Great Britain for the equivalent stages of production.

THE DISEASES OF ANIMALS (WASTE FOOD) ORDER 1973 (AS AMENDED) (SI 1936)

3.4.9 The majority of the articles in this Order entered into force in Great Britain on 1 July 1974 (the remainder earlier, in two stages). It revokes the Diseases of Animal (Waste Food) Order 1957. For the purpose of the Order, waste food means any waste material derived from livestock or poultry carcases, hatchery waste or eggs, and any foodstuffs which have been in contact with any of these. (It specifically excludes meal manufactured from animal protein, which is covered by the Processed Animal Protein Order 1989). The Diseases of Animals (Waste Food) Order prohibits the feeding to livestock or poultry of waste food, unless it has been processed under a licence granted by the appropriate Minister. The waste food must either be treated for at least 60 minutes at a temperature not less than 100°C or by another process authorised by the appropriate Minister. The Order also prohibits the possession of unprocessed waste food except by a licence holder. The Order is enforced by local authorities.

3.4.10 The equivalent Order in Northern Ireland is the Waste Food (Feeding to Livestock and Poultry) Order (Northern Ireland) 1974. Enforcement of the Order is undertaken by DANI.

ANNEX 3.5

STEERING GROUP ON THE MICROBIOLOGICAL SAFETY OF FOOD

The Steering Group was established in 1990, in response to a recommendation by the Committee on the Microbiological Safety of Food.

It manages surveillance and research and presents policy conclusions to Ministers. Its terms of reference are:

> "To identify through the surveillance the need for action to ensure the microbiological safety of food".

Members

Chairman

Dr J Steadman	Health Aspects of Environment and Food Division, Department of Health (DH)

Deputy Chairman

Dr H Denner	Chief Scientist (Food), Ministry of Agriculture, Fisheries and Food (MAFF)

Outside Experts

Professor J Arbuthnott	Principal and Vice-Chancellor, University of Strathclyde
Dr A Baird-Parker	Head of Microbiology, Unilever Research Ltd
Dr C Bartlett	Director, Communicable Disease Surveillance Centre
Dr E M Cooke	Deputy Director, Public Health Laboratory Service
Mr D Elliott	Technical Executive, Marks and Spencer plc
Mrs S Payne	Consumer interests
Mr E Ramsden	Chief Environmental Health Officer, Swansea
Dr T Roberts	Agricultural and Food Research Council, Institute of Food Research, Reading
Professor I K M Smith	Emeritus Professor of Microbiology and Parasitology, Royal Veterinary College, University of London
Mr M Verstringhe	Chairman of Catering and Allied Services Ltd

Officials

Dr R Cawthorne	Head of Animal Health (Zoonoses) Division, MAFF
Mr R Cunningham	Assistant Secretary, Health Aspects of Environment and Food Division, DH
Mr E Davison	Scottish Office Agriculture and Fisheries Department
Mr B Dickinson	Under Secretary, Food Safety Group, MAFF

Mr M Haddon	Under Secretary, Animal Health and Veterinary Group, MAFF
Mr I Henderson	Department of Agriculture for Northern Ireland
Mr E Kingcott	Chief Environmental Health Officer, Health Aspects of the Environment and Food Safety Division, DH
Dr A MacLeod	Scottish Office Home and Health Department
Dr E Mitchell	Department of Health and Social Services (Northern Ireland)
Miss D Pease	Under Secretary, Health Aspects of Environment and Food Division, DH
Dr R Skinner	Principal Medical Officer, Health Aspects of Environment and Food Division, DH

ANNEX 3.6

ANTHRAX

Reported cases of human and animal anthrax in Britain 1981-91

Year	Human Cases*	Livestock		
		Incidents	Deaths	
			Animals	No
1981	–	14	Cattle	17
1982	1	14	Cattle	17
			Pigs	1
1983	2	13	Cattle	13
			Pigs	1
1984	2	9	Cattle	9
1985	2	4	Cattle	6
1986	–	11	Cattle	12
			Sheep	2
			Horse	1
1987	1	6	Cattle	6
1988	2	3	Cattle	3
1989	1	1	Pigs	19
		3	Cattle	3
1990	3	5	Cattle	6
1991	1	2	Cattle	2

*No deaths reported
Sources: Human: reports to OPCS, HSE, CDSC
Animal: CVO reports (MAFF)

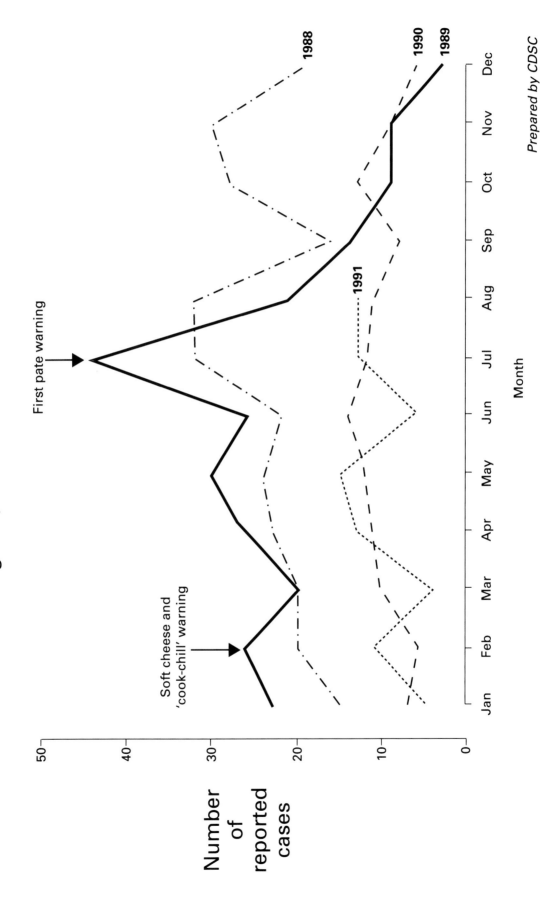

Human Listeriosis – Reports to CDSC and DMRQC
England, Wales and N. Ireland 1988-1991

Prepared by CDSC

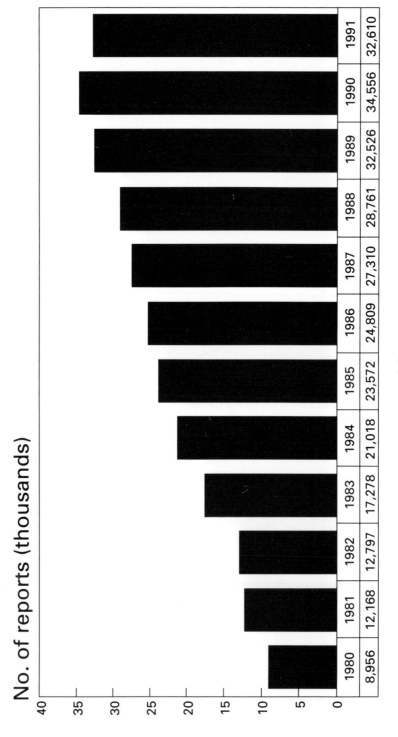

Campylobacter SP.
Laboratory Reports to CDSC*
England and Wales, 1980-1991

No. of reports (thousands)

Year	1980	1981	1982	1983	1984	1985	1986	1987	1988	1989	1990	1991
	8,956	12,168	12,797	17,278	21,018	23,572	24,809	27,310	28,761	32,526	34,556	32,610

Year

*Faecal isolates
1990 & 1991 data provisional

CHAPTER 4

MEDICATED FEEDINGSTUFFS

INTRODUCTION

4.1 This Chapter examines the mechanisms for licensing veterinary medicines, the conditions of sale for their incorporation into compound feed by both commercial compounders and on-farm mixers and the controls on incorporation. It also examines controls on the sale of final medicated feedingstuffs. Controls are applied at three stages; on premixes (which are licensed medicines), on intermediate feedingstuffs and on final feedingstuffs. All those who sell or supply these products are required to register their premises, and observe professional, or other, codes of practice. A full account of the legislation is in Annex 4.1.

SAFETY OF VETERINARY MEDICINES

4.2 The Medicines Act 1968 provides the basis for regulating the manufacture, sale and supply of all veterinary medicinal products in the UK. A central requirement of the Act is that all veterinary medicinal products must be licensed before they are marketed. The Veterinary Medicines Directorate (VMD), acting on behalf of the UK Agriculture and Health Ministers (who form the Licensing Authority) will issue a product licence only where the product, after a detailed scientific assessment, is found to be safe, effective and of sound quality. Where it is considered that an application for a product licence should be refused, the Act requires that advice should be sought from the Veterinary Products Committee (VPC).

Veterinary Products Committee

4.3 The VPC is an independent scientific committee established under Section 4 of the Medicines Act to advise the Licensing Authority on the licensing of veterinary medicines and to promote the collection of data on adverse reactions. Its members are experts in the variety of disciplines necessary to provide such advice and include practising veterinary surgeons. A list of members and terms of reference of the VPC is attached at Annex 4.2. The advice of the VPC is also always sought if the application involves an ingredient not previously licensed in the UK, an efficacy claim not previously accepted for a UK product licence, a new manufacturing process or route of administration, or if the product is in any other way novel. There are opportunities for appeal to the VPC and to the independent Medicines Commission where it is proposed that an application be refused. Neither the findings of the VPC nor of the Medicines Commission are made public although their views are reflected in the conditions attached to the product licence, and must by law appear on the product labels and in the product literature. Moreover, a list of the product licences granted is published in the London and Edinburgh Gazettes. **In the move towards greater openness about the activities of Government Committees we recommend that the VPC agenda and proceedings are made more freely available.**

CLASSIFICATION OF VETERINARY MEDICINES

4.4 Veterinary medicines for incorporation into feed fall principally into two main groups — Pharmacy and Merchants List (PML) products and Prescription Only Medicines (POM) products. A product is classified POM when the product is new or significantly alters the animal's physiology, has a POM classification in human medicine or when its use is judged to benefit from control by the veterinarian. Criteria for determining POM and PML classification are shown in detail at Annex 4.3. **We recommend that criteria for PML and POM classification be published as part of the regular reports of VPC activities.**

43

EC HARMONISATION OF CONTROLS ON VETERINARY MEDICINES

4.5 Medicines incorporated into feedingstuffs for therapeutic use are controlled under EC Directive 90/167 and are classified POM in the UK. Some antibiotics, growth promoters, and coccidiostats intended for long term application are licensed as PML products for incorporation into feed. These are listed in EC Directive 70/524. The procedures for harmonising controls on growth promoters and medicinal products are described in the following paragraphs.

4.6 Medicinal pre-mixes for use in feedingstuffs, like all other veterinary medicines, must be licensed before they are placed on the market. Licences are issued by regulatory authorities of individual Member States following the assessment of data supplied by applicants against the criteria of safety, quality and efficacy. EC Directives harmonise the data requirements, the nature of product literature, manufacturing requirements and basic controls over distribution. EC Directive 90/676 stipulates that products whose safe use requires the involvement of a qualified veterinary practitioner, are prescription only. At present, however, there are no harmonised Community lists of POMs and responsibility for implementing the criteria rests with national licensing authorities. **We have noted that in future a Member State may not authorise a new active ingredient for use in a veterinary medicine intended for food production animals until a Community-wide maximum residue limit (MRL) has been established or that it has been determined that an MRL is unnecessary.**

4.7 Companies wishing to secure licences in several Member States may ask for the data dossier to be considered by the Community's Committee for Veterinary Medicinal Products (CVMP) — a committee of officials drawing on their national sources of scientific expertise. To complete harmonisation in this sector the Commission has proposed changes which would have the effect of making CVMP opinions binding. The current proposals lack formal arrangements for the CVMP to have access to independent scientific advice. We think that the advice of independent experts is central to such an approval system. **We recommend that efforts continue to be made by the UK to secure the advice of independent experts in the structure of any European approval system for medicinal products.**

EC HARMONISATION OF CONTROLS ON PMLs IN FEEDINGSTUFFS

4.8 Harmonised lists of products under Directive 70/524 contain both medicinal and non medicinal additives as described in Chapter 5. Dossiers on both groups seeking listing are evaluated for quality, safety and efficacy by the EC Committee of Experts on Additives, Bioproteins and Undesirable Substances in Feedingstuffs. Adoption is by qualified majority voting. Some are referred for independent scientific advice to the Commission's Scientific Committee on Animal Nutrition (SCAN). Membership and terms of reference of this Committee is shown at Annex 5.4. Antibiotics, coccidiostats, and growth promoters currently listed in EC Directive 70/524 will at some time be moved across to EC Directive 81/851. They will then be dealt with in the Commission by the same Directorate as for veterinary medicines. There are also proposals to review safety, quality, and efficacy of medicinal additives approved many years ago to ensure that they meet current standards and to restrict new authorisations to ten years, renewable for further ten year periods.

SALE OF POMs AND PMLs

4.9 In principle the sale for administration of a POM in feed must be authorised by a veterinarian. A PML product for incorporation into animal feedingstuffs may be sold without prescription by pharmacists, compounders, veterinarians and registered distributors, to registered feed manufacturers. PML medicines not intended for incorporation may also be sold to farmers and other commercial keepers of animals for their use. PML products can only be sold from premises registered with the Royal Pharmaceutical Society of Great Britain (RPSGB) and in Northern Ireland with the Department of Health and Social Services (DHSS NI). Merchants who sell PML products are divided into two categories: Category 1 merchants may sell all PML products. Category 2 merchants may only sell a PML product for feed incorporation in the form of an intermediate medicated feed. Both categories are, however, subject to codes of practice and inspection by the registration or other appropriate authorities.

4.10 There has been some criticism of the arrangements for two separate categories of merchants (described in paragraph 4.9) whereby each meets different standards. We understand that an undertaking has been given to review the registration scheme in 1993.

4.11 The sale and supply of final medicated animal feedingstuffs including both POMs and PMLs are controlled by the Medicated Animal Feeding Stuffs Regulations 1992 which require that manufacturers who sell final medicated feed must be registered with the RPSGB or DHSS NI. Under the Medicines Act POMs, PMLs and feedingstuffs containing medicinal additives must carry a label which shows the name of the active ingredient, purpose for which it is intended, directions for use and withdrawal period. The observance of the withdrawal period has recently been made a statutory requirement.

MANUFACTURE OF MEDICATED FEEDINGSTUFFS

4.12 The incorporation of medicinal products into animal feed must only be undertaken by registered manufacturers. A register of manufacturers is maintained in England, Scotland and Wales by the RPSGB and in Northern Ireland by the Department of Agriculture (DANI). The Register is divided into two Categories — A and B. Category A lists feed manufacturers (mainly large commercial compounders) who are authorised to incorporate medicinal additives at any rate. Category B lists manufacturers (mainly on-farm mixers) who are authorised to incorporate medicinal additives at a rate above 2 kilograms per tonne. The distinction between incorporation rates above and below 2 kilogram recognises the fact that a medicinal additive is more evenly distributed when incorporated in quantities over 2 kilogram.

4.13 The Regulations prohibit a registered feed manufacturer from incorporating a medicinal product into an animal feed unless it is licensed for such incorporation in accordance with the product licence. A veterinarian may however prescribe, under the terms of a Veterinary Written Direction, an instruction to a registered manufacturer to incorporate a medicine outside the terms of its product licence, ie for use in a non-indicated species, or at increased dosages, but he can only do so if the product is licensed for incorporation in feed.

4.14 There was some evidence of confusion in the information submitted to us as to whether the generality of veterinary medicines could be used in feed. Some sectors of the industry were under the impression that, irrespective of licence conditions, all veterinary medicines could be used in feedingstuffs. This is not the case. Products can only be used in accordance with the product licence. **We recommend that it be made clear in the regulations and in guidance to the industry that veterinary medicines can only be used in feedingstuffs in accordance with the product licence.**

EXEMPTION FROM MEDICATED FEEDINGSTUFFS LEGISLATION

Fish farmers

4.15 Fish feed is subject to the medicated feed controls in that feed manufacturers who sell medicated feed to fish farmers must be registered with the RPSGB or DANI scheme. Fish farmers who mix their own feed are exempted from the requirement to so register. Fish farmers emphasised that fish diseases occur quickly and demand a swift response and treatment with a veterinary medicine or a medicated feed. We take the view that a requirement to register under the RPSGB scheme would not jeopardise the particular needs of fish farmers. We were impressed with the well-run establishments of farms visited and thought that they could probably meet registration standards. **We recommend that fish farmers who use medicated additives should be registered with the RPSGB or DANI.**

On-farm mixers

4.16 Some on-farm mixers are exempted from the requirement to register with the RPSGB or the DHSS NI because they apply medicated feedingstuffs either as a top dressing or mixed in water and neither of these methods are considered as falling within the current interpretation of "incorporation" in the legislation. We found it difficult to accept that mixing operations such as these, which may be carried out with less precision than those operations currently regarded as incorporation, were exempt from important controls by inspection. **We recommend that on-farm mixers using medicinal additives and intermediate medicated feedingstuffs in any manner should be registered with the RPSGB or DANI.**

4.17 There are a number of other exemptions including, for example, emergency treatments, mixing operations by someone who is not on either of the two RPSGB or DANI registers, or incorporation at a rate below 2 kilogrammes per tonne by someone who is only registered on list B. These treatments are always authorised by a veterinarian and we are satisfied with that arrangement.

POST LICENSING SURVEILLANCE

4.18 There are voluntary arrangements under the Suspected Adverse Reactions Surveillance Scheme whereby veterinary practitioners and farmers are encouraged to report any adverse reaction to veterinary medicinal products, including those used in feedingstuffs, to the VMD. Companies are required to make available to the VMD any such reactions reported to them. All cases are investigated and reported at three-monthly intervals to the VPC for its consideration and advice. Product licences can be amended if necessary, and procedures exist for the suspension or revocation of product licences where this is appropriate. The results are published in the VMD Annual Report. We are satisfied with these arrangements.

VETERINARY RESIDUES

4.19 The national programme of non-statutory Government surveillance is co-ordinated by the MAFF Working Party on Veterinary Residues in Animal Products (WPVR) which reports to the Steering Group on Chemical Aspects of Food Surveillance, a senior Government advisory committee (see Annex 5.3). This programme covers the national food supply by sampling UK-produced meat at abattoirs, imported meat at the port of entry into the UK, and meat at retail outlets. The WPVR also considers the results of statutory chemical analyses co-ordinated by the VMD under the National Surveillance Scheme for residues in meat. The results of all chemical analyses considered by the WPVR are published periodically in Food Surveillance Papers.

4.20 The WPVR has collated the results of the above surveillance programmes between May 1986 and December 1990. In this period over 13,300 chemical analyses were conducted for residues of antimicrobials, anthelmintics and coccidiostats in meat, liver, kidney, poultry, eggs and milk. Residues of lasalocid, a coccidiostat, were detected in one sample of chicken muscle and three samples of eggs from a total of 405 samples analysed; no other coccidiostats were detected. Residues of oxfendazole, a benzimidazole anthelmintic agent, were detected above the statutory maximum limit in one sample of sheep liver from a total of 809 analysed for benzimidazoles. No residues of other anthelmintic agents were detected. At the time this surveillance was undertaken, maximum residue limits (MRLs) had been recommended by the VPC for a range of veterinary medicines. Residues of tetracyclines, a class of antimicrobial compounds, were detected above these MRLs in 18 of 404 samples of UK sourced kidneys and in 13 of 202 samples of imported kidneys. These results also included over 4,900 analyses for residues of sulphonamides in pig tissues; the incidence of residues above the recommended MRLs in kidneys from pigs produced in Great Britain has declined from 15% in 1986 to 5% in 1990 and to 3% in 1991. The results of surveillance for sulphonamides in 1991 will shortly be published in the Annual Report of the VMD. **We note that despite considerable effort by all concerned sulphonamide residues persist and we recommend that careful consideration be given to the licensing of drugs for in-feed use which require long withdrawal periods.**

4.21 The WPVR has also reported that the incidence of residues of synthetic hormones has declined from 0.7% in 1988 to 0.3% in 1990, confirming compliance with the legal ban on their use. No residues of non-hormonal growth promoters or tranquillisers were detected in surveillance programmes between May 1986 and December 1990.

4.22 The WPVR also co-ordinates the surveillance of veterinary residues in farmed fish. Between 1987 and 1989, oxytetracycline was detected in 11% of farmed salmon and trout samples. No samples exceeded the MRL for oxytetracycline recommended by the VPC at that time. Oxolinic acid was detected in 22% of samples of farmed salmon and trout analysed between 1987 and 1989. As a result, the VPC extended the withdrawal period for this antimicrobial agent. Only one out of 49 samples of salmon analysed between October 1989 and July 1990 contained a detectable residue of oxolinic acid. No residues of other veterinary medicines were detected in farmed fish. It is anticipated that in future there will be an EC obligation to survey farmed fish, and also poultry and game.

4.23 These findings, which are published in the WPVR Report, have been assessed by the various Committees which have responsibilities on these matters, namely the Committee on Toxicity of Chemicals in Food, Consumer Products and the Environment, the Food Advisory Committee and the Veterinary Products Committee. **Their conclusions are that the Report provides reassurance that, in general, foods of animal origin contain little or no residues of veterinary products and that where present they pose no significant risk to human health. They recommended that surveillance should be continued and should be increased in certain areas (poultry, eggs, fish). We accept their findings and recommendations subject to the reservation expressed in paragraph 4.20.**

NEW CONTROLS ON VETERINARY RESIDUES

4.24 During our review new regulations were introduced which strengthen the existing controls on veterinary residues. The new measures make it an offence to use unauthorised drugs and to sell animals or meat containing residues in excess of MRLs. They require farmers to observe specified withdrawal periods. They enable the detention of animals and meat where misuse is suspected. These will have an obvious impact on the livestock farmer. If, as is intended, all carcases at abattoirs can ultimately be traced back to the primary producer, this legislation will make it necessary for livestock farmers to be scrupulous in their feeding practices and, if they are on-farm mixers, to take care in the way they incorporate or otherwise mix medicated additives and intermediate medicated feedingstuffs. The changes should further reduce the incidence of residues enumerated in paragraph 4.20 ad 4.21 above. **We wish to reaffirm that these legislative developments do not remove the need for the registration and inspection of farm and compounders premises where medicinal additives and intermediate medicated feed are used.**

ANTIBIOTIC RESISTANCE

4.25 Approaches to the use of antibiotics in animal husbandry were defined in the (Report of the Joint Committee on the use of Antibiotics in Animal Husbandry and Veterinary Medicine Swann report) where it was concluded that the use of antibiotics as growth promoters was a major contributory factor to the appearance of multiple antibiotic resistance in zoonotic salmonella. The report recommended that antibiotics used in human medicine should not be used as growth promoters in animals. Advice on antibiotic usage in animals is given by the VPC and the structure and functions of that committee have been described in Annex 4.2. In relation to antibiotic usage the VPC considers the consequences of antibiotic residues as well as antibiotic resistance in bacteria. Antibiotic residues in food have already been considered in paragraphs 4.20 to 4.24. The Expert Group heard information from the Public Health Laboratory Service (PHLS) on the antibiotic resistance of salmonella. In chapter 3 we have discussed the sources of human salmonella and have cited the views of the Committee on the Microbiological Safety of Food that "salmonellosis in man is usually foodborne from infected food animals."

4.26 The PHLS data for the 10 year period from 1981 showed that the incidence of resistance in human isolates of Salmonella enteritidis has remained essentially unchanged. As this organism is at the present time the commonest cause of human foodborne illness and as non-invasive human salmonellosis does not require antibiotic therapy there may appear to be no immediate cause for concern with this serotype. However, we have received information which underlines the need for caution. Some human infections with salmonella, particularly septicaemias, do require antibiotic therapy. In addition we have been told that the availability of new effective antibiotics means that human disease which was in the past not treated with antibiotics is now more frequently so treated. *S. typhimurium* from cattle have become increasingly multiple antibiotic resistant; in 1981 15% of such strains were multiple antibiotic resistant whereas in 1990 the figure was 66%. Comparable figures for *S. typhimurium* from poultry were 2% and 8%. This suggests that the use of antibiotics for prophylaxis and therapy, at least in the cattle industry, has contributed to the emergence of multiresistant strains. We were also told that decreased sensitivity to nitrofuran drugs and resistance to nalidixic acid are beginning to increase in strains of *S. enteritidis* and *S. virchow* from poultry.

4.27 The use of a novel aminoglycoside apramycin in bovines has allowed for the first time an estimate to be made of antibiotic resistance due to antibiotic usage in animals. Although there is cross resistance between gentamicin (an antibiotic important in human medicine) and apramycin such cross resistance is not absolute and different mechanisms of resistance can be distinguished. We have received evidence that gentamicin resistance in *S. typhimurium* type 204c strains has arisen because of the use of apramycin in bovine husbandry and that 7 of 26 (27%) gentamicin resistant human clinical isolates of *E. coli* submitted for examination to the PHLS Antibiotic Reference Laboratory were also resistant to apramycin. Antibiotic resistance in *E. coli* is extremely important as the organism is a significant cause of human disease, particularly urinary tract infection. **We have referred both issues, ie increasing multiple antibiotic resistance of salmonella and apramycin/gentamicin resistance, to the VPC for consideration. We have been told that they consider that the prophylactic use of antibiotics with cross-resistance to those used in human medicine should be strongly discouraged (see Annex 4.5). We welcome this view. We recommend that monitoring of antibiotic resistance of salmonella be continued and that routine monitoring of antibiotic resistance in human E. coli be established. The possible contribution of antibiotic usage in animals to antibiotic resistance in human isolates should, where feasible (ie as with apramycin), be assessed. We also recommend that not only should antibiotics giving cross-resistance to those used in human medicine not be used as growth promoters but that their prophylactic use in animals be reconsidered.**

ENFORCEMENT OF MEDICATED FEEDINGSTUFFS LEGISLATION

4.28 Several bodies have a role in the enforcement of medicated feedingstuffs legislation. Local authorities enforce labelling provisions. The RPSGB and DANI check compliance with the registration conditions. Local Authority Coordinating body on Trading Standards (LACOTS) reports that there is no provision in medicated feedingstuffs legislation to take a case against a compounder who declares the presence of a medicated additive on the feed label for which no trace can be found on analysis. We considered that this situation does not pose a risk to human, or animal health and can be satisfactorily dealt with under other 'consumer' legislation.

4.29 The RPSGB or DANI has responsibility for inspecting all registered premises where medicinal additives are incorporated into feedingstuffs to check that the necessary records and equipment are maintained and that there is compliance with the codes of practice. The numbers on the Category B register are falling. This could be for a number of reasons including a reluctance to pay the registration fee, less medication being used, or an increase in the use of top dressings and water treatments which do not require registration because these treatments are not regarded as "incorporation" (see paragraph 4.16). We have already recommended that these exemptions should not continue.

ANNEX 4.1

LEGISLATION: MEDICATED FEEDINGSTUFFS

MEDICINES ACT 1968

4.1.1 The Act provides the basis for regulating the manufacture, sale and supply of all veterinary medicinal products in the UK. The Act's definition of a medicinal product encompasses all products which affect the animal's physiology irrespective of whether they are used to cure or prevent disease. The Act also requires that all veterinary medicinal products placed on the market should be licensed. The Veterinary Products Committee was set up under Section 4 of the Act.

THE MEDICINES (MEDICATED ANIMAL FEEDINGSTUFFS) REGULATIONS 1992

4.1.2 The Regulations require that the incorporation of medicinal products should only be undertaken by registered manufacturers. They prohibit the sale, supply, or import of animal feedingstuffs containing a Pharmacy and Merchants List product unless that medicinal product has been incorporated in accordance with a product licence, animal test certificate or veterinary written direction; or from selling or supplying an animal feedingstuff containing a Prescription Only Medicine (POM) unless it has been incorporated in accordance with a veterinary written direction. A veterinary written direction is not required where incorporation of a POM into an intermediate feedingstuff is in accordance with a product licence or animal test certificate and the finished product is held by a category A manufacturer prior to sale or supply to another registered Category A manufacturer, neither of whom has animals under his direct control.

THE MEDICINES (VETERINARY DRUGS) (PHARMACY AND MERCHANTS LIST) ORDER 1992

4.1.3 Under this Order agricultural merchants who are not manufacturers of feedingstuffs but who sell intermediate medicated feedingstuffs are required to register their premises with the Royal Pharmaceutical Society of Great Britain; in Northern Ireland they are required to Register with the Department of Health and Social Services for Northern Ireland. Registration carries with it the obligation to comply with the standards of conduct contained in the statutory codes of practice.

CODES OF PRACTICE

4.1.4 There are four statutory codes of practice covering Category A and B manufacturers and Category 1 and 2 distributors. These set out the standards of conduct which manufacturers and distributors are required to meet in order to comply with the provisions set out in legislation made under the Medicines Act 1968. These codes are described in paragraphs 4.5 and 4.8 of Chapter 4. Compliance with the codes is a statutory requirement if a manufacturer or distributor is to qualify for and retain registration.

ANNEX 4.2

VETERINARY PRODUCTS COMMITTEE

The Veterinary Products Committee was established in 1970 under the Medicines Act 1968 and gives independent advice to the Licensing Authority (ie Agriculture and Health Ministers) on the safety, quality and efficacy of veterinary medicinal products and on suspected adverse reactions to such products. Its full terms of reference are:

(i) to give advice with respect to safety, quality and efficacy in relation to the veterinary use of any substance or article (not being an instrument, apparatus or appliance) to which any provision of the Medicines Act 1968 is applicable;

(ii) to promote the collection and investigation of information relating to adverse reactions for the purpose of enabling such advice to be given.

Members

Chairman

Professor J Armour	Vice Principal, University of Glasgow
Professor J P Arbuthnott	Principal and Vice-Chancellor, University of Strathclyde
Dr D N Bateman	Medical Director, Northern Regional Drug and Therapeutic Centre, Newcastle-upon-Tyne
Professor P M Biggs	Visiting Professor in Microbiology, Royal Veterinary College, University of London
Professor J W Bridges	Professor of Toxicology and Director of the Robens Institute of Industrial and Environmental Health and Safety, University of Surrey
Professor J Brown	Professor of Pharmaceutical Chemistry, Sunderland Polytechnic
Dr D H Calam (occasional member)	Head of Chemistry Division, National Institute for Biological Standards and Control, Hertfordshire
Mr D. S. Collins	Veterinarian, Consultant in Veterinary Public Health, Belfast
Dr A Cooke	Head of Pollution Branch of Science Directorate, English Nature
Mr P Crossman	Practising Veterinary Surgeon, Chichester, West Sussex
Professor S P Denyer	Professor of Pharmacy, Head of the Department of Pharmacy, Brighton Polytechnic
Miss K Gibson	Practising Veterinary Surgeon, Edinburgh
Dr R J Heitzman	Private consultant working for Food and Agriculture Organisation/World Health Organisation, European Community, International Atomic Energy Agency and United Nations Development Programme
Professor D E Jacobs	Professor of Veterinary Parasitology, Department of Pathology, Royal Veterinary College, University of London
Professor G E Lamming	Professor of Animal Physiology, Department of Physiology and Environmental Science, University of Nottingham

Professor P Lees	Head of Department of Veterinary Basic Sciences, Royal Veterinary College, University of London
Dr K A Linklater	Director of Veterinary Investigation Services, Scottish Agricultural College, Edinburgh
Professor R Richards	Deputy Director, Institute of Aquaculture, Stirling University
Professor I K M Smith	Emeritus Professor of Microbiology and Parasitology, Royal Veterinary College, University of London
Dr S Venitt	Team Leader, Molecular Carcinogenesis Section, Institute of Cancer Research

ANNEX 4.3

VETERINARY PRODUCTS COMMITTEE GUIDELINES FOR THE CLASSIFICATION OF PRODUCTS "POM" AND "PML"

Products indicated for use in farm animals will normally be classified as P [ie can only be sold by a pharmacy] unless a more restrictive category is appropriate or it can with safety be classified as PML. In deciding whether to advise the licensing authority to classify a product as POM or PML, the Veterinary Products Committee will be guided by the following criteria.

PRESCRIPTION ONLY MEDICINES (POM)

The POM category may be appropriate when one or more of the following apply:

(i) when diagnosis of the condition for which the product is intended would be beyond the competence of the livestock owner or accurate diagnosis (including in particular differential diagnosis) is required so that the medication appropriate to the circumstances can be administered and the necessary veterinary advice given;

(ii) when the product needs to be administered by the veterinary surgeon in person or under his supervision, for example, because it is a therapeutic antimicrobial agent or an injectable preparation for small animals; or because of other legal provisions; or when the method of administration is novel;

(iii) when the drug's toxicity may present a safety hazard in animals or man; when the risk/benefit ratio of using the product is finely balanced; when the substance has significant activity on the central nervous systems (eg anaesthetics, tranquillisers) or where it significantly alters the animal's physiology;

(iv) when careful monitoring of the use of the product is required; this would normally apply to all new active ingredients for up to five years and could apply to known active ingredients licensed for administration by a new route;

(v) when the substance is controlled under the Misuse of Drugs Act or allied legislation, or the active ingredient is classified as POM for human use.

PHARMACY AND MERCHANTS LIST (PML)

Products for use in pet animals, racing pigeons, ornamental fish and other non-food species will normally be P unless the product is suitable for general sale or one or more of the guidelines set out above apply, thus making a POM classification necessary. In the case of products indicated for use in farm animals, products which do not need a POM classification will be P unless none of the conditions below apply, when the product may be classified as PML.

(i) when advice is needed on potential risks to the person administering the product;

(ii) when advice is needed on the possibility of undesirable interaction with other widely used veterinary drugs; when the user would benefit from point-of-sale advice which may include referral to a veterinary surgeon;

(iii) when point-of-sale advice on the method of use or the preparation of a product prior to use might be required:

(iv) where usual storage conditions (including storage during usage) or unusual requirements for safe disposal should be brought to the user's attention.

ANNEX 4.4

THE COMMITTEE FOR VETERINARY MEDICINAL PRODUCTS

COMPOSITION

One representative from each Member State, usually, national officials responsible for advising on the authorisation of veterinary medicinal products.

One representative of the Commission.

TASKS

To try to ensure the adoption of common positions by Member States on decisions relating to the authorisation of veterinary medicinal products, including:

— applications for authorisation submitted in accordance with Community procedures (Directives 81/851/EEC and 87/22/EEC);

— advice on any other question relating to the authorisation of veterinary medicines raised by a Member State for the Commission.

To advise the Commission on any detailed changes which may be necessary to the Community legislation on the testing of veterinary medicinal products.

WORKING PARTIES OF THE CVMP

1. Safety of residues; toxicological assessment of the safety of individual compounds used in food-producing animals; general advice on any question relating to the safety of residues.

2. Efficacy of veterinary medicines; preparation of guidelines on the conduct of efficacy studies on veterinary medicines.

3. Immunological veterinary medicinal products:

— technical advice on applications for rDNA vaccines submitted in accordance with Directive 87/22/EEC;

— preparation of detailed technical standards for immunological veterinary medicinal products.

4. *Ad hoc* group on hormones; advising on the continued use of hormonal compounds for therapeutic or zootechnical purposes.

5. *Ad hoc* group on pharmacovigilance: to develop and coordinate an EC Pharmacovigilance Scheme.

Veterinary Products Committee

Veterinary Medicines Directorate
Woodham Lane, New Haw, Weybridge, Surrey KT15 3NB
Telephone: 0932-336911 ext. **3050**   Telex: 262318   Fax: 0932-336618

14 April 1992

Mrs. E. Owen
Secretary
Expert Group on Animal Feedingstuffs
MAFF
Room 303c, Ergon House
17 Smith Square
London   SW1P 3JR

Dear Mrs. Owen

ANTIBIOTIC RESISTANCE

The Veterinary Products Committee has now had the opportunity to discuss certain aspects of antibiotic resistance in salmonellae and apramycin/gentamicin resistance.

The Committee noted that in man in the UK *S. enteritidis* was currently the most prevalent but least resistant serotype, whereas *S. typhimurium* and *S. virchow* were less prevalent but more resistant. *E. coli* could be a reservoir of plasmids for salmonellae and the Committee suggest that it might be included in the monitoring programme.

The Committee noted that the epidemiological evidence demonstrates that in the UK, people mostly become infected with salmonellae from the food chain. Although there is a high level of resistance from the over use of antibiotics in the Middle and Far East, less than 10% of UK human infections are in travellers from abroad. Approximately 90% of human infections in the UK originate in the food chain in the UK, largely from the food itself rather than from the food handler. Cattle, pigs and poultry have been identified as the main sources.

Use of antibiotics against bacterial infections in both animals and man can result in resistance, but the Committee was unaware of evidence showing that for many antibiotics their use in animals significantly compromised their corresponding use in people. The risk of contracting resistant bacteria from fish - in which antibiotics are used to control some bacterial diseases - was thought to be minimal at any rate in the meantime. The use of antibiotics in people, including for the treatment of diseases other than those caused by salmonellae, could nevertheless lead to resistance in salmonellae.

Continued

The outcome of the use of any antibiotic in any population at large is unpredictable; hence the application of all antibiotics must always be cautious. Antibiotic use should not be indiscriminate even for conditions affecting human health and there are many clinical situations in man where antibiotics are not appropriate. Widespread use of antibiotics in human medicine unquestionably leads to the emergence of resistant strains of bacteria. The availability of new, potent antibiotics means that some enteric infections, for which in the past antibiotic use had been discouraged, are now treated with antibiotics, particularly when the cases are so severe as to require hospital admission. The Committee considered that such drugs should not necessarily be precluded from licensing for therapeutic use in animals but that licensing for prophylactic use was unacceptable and such use should be strongly discouraged. The Committee felt that it should consider each instance involving a compound new to veterinary medicine on the merits of the particular case and seek appropriate additional advice where necessary.

A copy of this letter goes to Professor Lamming and Professor Armour.

The Committee will keep this situation under review.

Yours sincerely

J. P. O'Brien
Secretary

CHAPTER 5

FEED COMPOSITION AND MARKETING

INTRODUCTION

5.1 The manufacture and supply of animal feedingstuffs has been subject to UK law since the last century. The primary legislation underpinning the detailed controls is the Agriculture Act 1970 (Part IV), as amended. This was based on the 1926 Fertiliser and Feedingstuffs Act but brought it up-to-date and introduced additional measures. Its principal requirements are that feedingstuffs, when sold, should be fit for their intended purpose and free from harmful ingredients. A statutory statement is required as to the composition of the feed and information and instructions must be provided on storage, handling and use.

5.2 There are two sets of secondary regulations on animal feedingstuffs made under Part IV of the 1970 Act. The first is the Feeding Stuffs (Sampling and Analysis) Regulations 1982 which is dealt with in Chapter 6. The second is the Feeding Stuffs Regulations 1991. The latter Regulation requires that a statutory statement accompanies all feedingstuffs. The statement can include nutritional information, such as amounts of protein, oil, fibre and ash; names or categories of ingredients. The Regulations also contain precise definitions of certain materials used in feedingstuffs; limits of variation that apply to the various declarations; the permitted additives and conditions for their use; maximum limits for listed undesirable substances; detailed requirements relating to specified protein sources, and labelling requirements for additives and pre-mixtures. UK and EC Committees are closely involved in developing feedingstuffs policy and legislation. A fuller account of the legislation is given in Annex 5.1.

5.3 Although the scope of the Agriculture Act and the Feeding Stuffs Regulations made under it is extensive, we found some weaknesses, both legislative and administrative, which should be addressed. These are discussed in paragraphs 5.8-10, 5.14, 5.22 and 5.24.

CONTROLS ON FEED INTENDED FOR SALE

Compound feed labelling

5.4 All feedingstuffs must be clearly labelled. Types of feedingstuffs range from straight feedingstuffs, which are generally made from one ingredient, to compound feedingstuffs, which include several. The requirements for compounds vary depending on whether the feed is complete or complementary, also according to the species of animal for which it is intended. The arrangements for compound feedingstuffs are now largely harmonised throughout the EC. They are implemented in Great Britain in the Feeding Stuffs Regulations 1991 which came into force on 22 January 1992. Manufacturers must declare specified information on the label and can elect to declare certain further information. Labels must give information about the type of product, the animal species for which it is intended and the feeding rates; they must also show the presence and levels of components such as oil, protein, fibre, ash, moisture, and technological additives, such as emulsifiers, colours and antioxidants. Feed ingredients must be declared. The manufacturer can choose to declare by categories or individually. If categories of ingredients are chosen they must comply with the EC list of categories included in the Feeding Stuffs Regulations 1991. If names of ingredients are given in full they will, in future, have to comply with EC nomenclature. There are additional options to declare energy values, certain amino acids and other constituents according to their importance to different species of animal.

Labelling of straight feedingstuffs

5.5 Straight feedingstuffs are normally supplied to farmers primarily by merchants and by companies who are also compounders, and in many instances the same materials are used as raw materials for compounded feeds. There is generally no obvious difference between a material which can be sold as a straight feedingstuff or as a raw material. The EC controls on straight feedingstuffs are much less detailed than those for compounds and those for raw materials are less demanding than those for straight feedingstuffs. An artificial distinction is created because the seller decides to describe and label a material in a particular way. We formed the view that materials which could be regarded equally as raw materials or straight feedingstuffs tended to be sold as raw materials thus obviating the need for some labelling provisions and compliance with some statutory limits for contaminants.

5.6 The EC Commission is considering extending the standards for contaminants in raw materials and is likely to propose a review of the EC labelling provisions for straights. **We recommend that during any review by the Commission of the provisions on straights and raw materials that the opportunity should be taken to rationalise the marketing and labelling arrangements for these two types of feedingstuffs.**

Labelling of feed additives

5.7 Feed additives include medicinal substances, (which are discussed earlier in Chapter 4), vitamins, trace elements and substances with a technological function such as stabilisers and binders. Vitamins and technological additives can be sold as single substances or premixtures which are mixtures of additives often with a carrier. The labels for single additives must show the content of active substance and may declare the EC number and directions for use, including any safety recommendations. The labels for premixtures must declare the level and names of the additives, species or category of animal for which the premixture is intended and directions for use. Vitamins have an extra labelling requirement because they have a limited shelf life, consequently the period for which they remain effective must be shown on the product label.

CONTROL OF FEED NOT INTENDED FOR SALE

5.8 Feed produced on farms accounts for about 40% of the total feed consumed by livestock but is not subject to the full labelling and compositional provisions of the Feeding Stuffs Regulations 1991. It is, however, required to comply with the contaminant limits, the ban on specified ingredients, and the use of permitted additives as laid down in the Regulations. **That farm-mixed feeds should meet the contaminant and other regulations that are designed to protect human and animal health is a fundamental requirement and is clearly the intention of the Regulations. However, there are presently no systematic arrangements for ensuring compliance with the Regulations. We therefore recommend that farm-mixed feed should be sampled and inspected to ensure that it complies with the statutory limits for undesirable substances and additives in feedingstuffs. There is need to strengthen powers of entry for enforcement officers for this purpose. Some on-farm mixers and integrated units may not be readily identifiable. We recommend that they be identified and listed for inspection purposes.**

Identification of on-farm mixers, and integrated units

5.9 Some farm businesses are identified and inspected. The RPSGB or DANI inspect manufacturers using medicinal additives. Also some farm businesses are registered with MAFF as approved buyers of certain contaminated raw materials because they have mixing and storage facilities which can guarantee their ability to produce a final feed which meets statutory requirements. Unlike farm businesses registered with the RPSGB they are not inspected and should be. **We recommend that there should be powers to check that these businesses have the proper facilities and are able to deal with contaminated materials satisfactorily.**

5.10 Evidence presented to us showed that some farmers are not well informed about their current responsibilities under the Feeding Stuffs Regulations 1991. It is important that they should be familiar with, and conform to, the standards for undesirable substances in feedingstuffs and the list of banned feed ingredients. On-farm mixers and integrated units should be encouraged to adopt the general principles of good manufacturing practice and hazard analysis and critical control points (HACCP). Guidance on this could usefully be incorporated in a code of practice. **We recommend that Government provide a code on the principles of HACCP and good manufacturing practice explaining the responsibilities all feed manufacturers have under the Feeding Stuffs Regulations 1991.**

EVALUATION OF FEED INGREDIENTS

5.11 New feed ingredients, or ingredients produced by new processing technology, are subject to the general safety provisions of the Agriculture Act but there is no statutory requirement for their initial evaluation for safety or nutritional quality. The exceptions are bioproteins such as yeasts, bacteria and algae, and, as regards safety, products containing genetically modified organisms which come under the scrutiny of both UK and EC expert committees. The EC Commission has indicated that it will review and extend the controls on bioproteins to include protein sources which are products of traditional processes.

5.12 We have considered whether all feed ingredients and manufacturing processes should be evaluated and approved or whether all feed ingredients should require certification on quality and efficacy criteria. We are conscious that changes in the production and use of animal protein in feed are generally agreed to have been significant in the development of BSE in the UK, and that approval systems are operated in some countries, and these report no great difficulty with the arrangement. However, we have concluded that a universal system of prior approval would place an unnecessary burden on feed manufacturers and would inhibit innovation and the development of new technology. We consider that the best approach is to provide a method under which in the introduction of new materials, or significant changes in processes or manufacturing techniques are kept under review by independent experts. **We recommend that in future an independent Animal Feedingstuffs Advisory Committee should keep a watching brief on feed developments and keep under review data on feed composition and nutritional needs of livestock so as to maintain the safety of the food chain.** We have commented at paragraph 5.29 and in Chapter 7 on the form and remit of such a committee.

EVALUATION OF FEED ADDITIVES

5.13 There is an EC approved list for feed additives in Directive 70/524. In general, other additives cannot be used. Those additives which are permitted for use throughout the Community will have been considered in the UK for their safety, quality and efficacy. The first stage of EC approval of an additive, once safety issues have been considered, usually allows Member States to choose whether or not to allow it to be marketed on their territory. The second stage occurs when the remaining data on quality and efficacy satisfies the EC Committee on Additives, Bioproteins and Undesirable Substances, and all Member States must then allow the additive to be marketed in their territory.

5.14 National rules are deemed to apply to classes of additives (silage agents, probiotics and enzymes) which have yet to be approved and are currently outside EC arrangements. In the UK the general provisions of the Agriculture Act apply but there are no specific legal powers to prevent the marketing of the additives, even though their safety has not been evaluated. **These classes of additives are due to be assessed through the EC expert committee system and the EC should be encouraged to start this process. However, it is recommended that, if there is delay in the EC consideration, UK experts should begin to evaluate the products on the market in the UK. We also recommend that any new class of additives, not covered by EC controls, which appear on the UK market should be referred for independent scientific advice. There are no powers to require manufacturers to notify Government of new additives and we recommend that there should be powers to require them to do so.**

PROBLEMS INVOLVING ADDITIVES

5.15 Because of the harmonisation of additives legislation across the EC there are potential constraints on the speed at which emerging problems involving additives might be resolved. Recent developments provide appropriate illustration of the present arrangements at work.

Vitamin A

5.16 Evidence presented to a Council of Europe meeting in 1990 indicated that pregnant women could be exposed to a risk if they consumed high levels of Vitamin A. The UK Chief Medical Officer advised pregnant women and women who might become pregnant not to take Vitamin A supplements except under medical supervision and not to eat liver or liver products because high levels of Vitamin A had been found in samples of liver. Since Vitamin A is frequently included in feed by compounders, UK advisory committees moved quickly to review the typical level of use of Vitamin A in feedingstuffs and reached agreement with industry on lower inclusion rates. These proposals were considered by the appropriate EC committees which within three months introduced new Community-wide statutory levels for Vitamin A inclusions in livestock diets.

Canthaxanthin

5.17 Canthaxanthin is used in fish feed to colour fish flesh and in poultry feed to colour egg yolk. Examination of new data led UK food and toxicity advisory committees to conclude that the use of canthaxanthin should cease. The data indicated that canthaxanthin is deposited in the human retina with the possibility that long term intake could result in changes of vision in some individuals. Ministers have accepted this advice and urged a ban in the Community. However, EC scientific committees have concluded that a ban would not be appropriate and that action should await the publication of further toxicological data in 1992.

5.18 There is clearly some scope for differences in view between UK and EC committees which may lead to a different degree of priority being given to the restrictions on the use of particular feed additives. However, we do not believe that this represents a major area of concern or one which requires the introduction of powers beyond those presently available to the UK government.

UNDESIRABLE SUBSTANCES IN FEEDINGSTUFFS

5.19 The Feeding Stuffs Regulations 1991 contain a list of ingredients which are banned for use in compound feedingstuffs. They include faeces, urine, sawdust, leather and sewage sludge. Other banned ingredients are discussed in Chapter 2. **We recommend that the list of banned ingredients should be kept under review by the Animal Feedingstuffs Advisory Committee.**

5.20 There are statutory limits for certain contaminants in feedingstuffs including organochlorine pesticides, heavy metals, and aflatoxin. An EC proposal is likely to require consignments of raw materials to be clearly marked when contamination with, for example, heavy metals exceeds specified levels. In the case of pesticides, statutory limits (under pesticides legislation) will, in future, be set for residues in crops irrespective of whether the crop is destined for human or animal consumption. **These measures increase the degree of protection afforded to commercial compounders and on-farm mixers and are to be welcomed.**

INCIDENTS OF FEED CONTAMINATION

Lead incident

5.21 There are few reported feed contamination incidents. Those that are reported can generally be attributed to human error and a failure of quality control systems. One significant episode in 1989/90 involved imported maize gluten replacer pellets highly contaminated with lead. Supplies reached compounders and farms in the Netherlands and UK when a cargo scheduled for destruction was fraudulently diverted into the feed chain. The incident was instrumental in influencing the EC Commission to produce proposals to strengthen notification arrangements which alert Member States to consignments of contaminated feed and to propose labelling of contaminants in raw materials in order to harmonise controls on raw materials in the different Member States. It also highlighted the inadequacy of manufacturers' record keeping which lead to difficulties in tracing purchasers of contaminated feed. **We accept that it is impossible to envisage legislative controls which are proof against fraudulent activity. We endorse the measures that are being promulgated to limit contamination and reduce potential risks and we** recommend that arrangements are made to ensure that all manufacturers who sell or supply feed should have detailed records which will allow feed purchasers to be identified.

5.22 There are powers under the UK Food and Environmental Protection Act 1985 to intercept contaminated supplies of feedingstuffs when there is a risk to human health, and these were used during the incident involving lead contamination. However, we have been advised that neither those powers nor any available under the Agriculture Act are adequate to allow routine inspection of feedingstuffs at ports. We believe this is an important requirement especially when the UK is the point of entry to the EC. **We recommend that powers should be available for the routine inspection of feedingstuffs and raw materials imported from non-EEC countries into the Community via the UK.**

FEED SURVEILLANCE ARRANGEMENTS

5.23 There are no routine, coordinated feed surveillance arrangements to assess chemical contamination of feedingstuffs in the UK, except for the limited work carried out under the auspices of the Local Authorities Coordinating Body on Trading Standards (LACOTS). This is designed to check compliance with statutory obligations. Results are not formally reported to central government, but are often made available on an informal basis. Some *ad hoc* surveys are undertaken by central government to gather information on levels of undesirable substances or microbiological contamination in feed and feed ingredients. These have assisted in discussions on setting maximum permitted limits in feed for a range of naturally occurring toxic substances and adventitious contaminants. There are limits for heavy metals, organochlorine pesticides, aflatoxin and certain plant material (referred to in paragraph 5.20).

5.24 The evidence from the very limited survey data available for residues of certain pesticides and aflatoxin in feed has not shown any significant areas of concern. There is continuing low level contamination within statutory limits. Nonetheless, it remains that there is no national forum for commissioning feed surveys or evaluating the results. We think it is essential for feedingstuffs to be regarded as part of the total food chain to allow the contribution feed may make to food contamination to be assessed. This would ensure that feed surveys are designed and undertaken in the most effective way and provide valid information about the quality and safety of feedingstuffs. In particular it would reinforce both industrial and consumer confidence. **We recommend that feed surveillance be formally integrated into the existing MAFF food surveillance arrangements which are carried out under the auspices of the Steering Group on Chemical Aspects of Food Surveillance. The results should be published.** The membership and terms of reference of the Steering Group on the Chemical Aspects of Food Surveillance is in Annex 5.3.

5.25 We have separately recommended in Chapter 3 that surveillance of feed for microbiological contamination could be integrated with food surveys proposed by Steering Group on Microbiological Safety of Food. **We recommend that all survey results be reported to the Animal Feedingstuffs Advisory Committee.**

EXISTING COMMITTEE STRUCTURE

UK Interdepartmental Committee on Animal Feedingstuffs

5.26 A number of committees in UK and EC advise on policy and regulatory matters. In the UK an interdepartmental committee in MAFF (which includes representatives from MAFF, Department of Health, VMD, Scottish and Northern Ireland Agriculture Departments) evaluates non-medicinal feed additives and bioproteins. It also considers ingredients and undesirable substances and provides a coordinated view on policy to assist in the formulation of national controls and in discussions and negotiations in the EC. The Committee has currently three sub-committees which examine the areas of research and development, probiotics, and feed ingredients respectively. A fuller account of the functions and membership is attached at Annex 5.2.

EC legislative committees

5.27 The number and composition of EC committees varies according to the number and subject of the directives and regulations under discussion. Normally there is an expert working group which discusses additives, undesirable substances and bioprotein controls and one which discusses compositional and marketing issues. These groups meet under Commission or Council auspices according to the status of the draft instrument under discussion. A description of the types of Community legislation and the process of agreement is at Appendix I.

EC advisory committee

5.28 The EC Commission has its own advisory committee — the Scientific Committee for Animal Nutrition. This is a committee of independent experts brought together to advise the Commission on a range of issues including the composition of feedingstuffs, processes, such as heat treatments, which modify feedingstuffs, and additives and contaminants in feedingstuffs. The conditions under which this Committee was established and its terms of reference and membership, are attached at Annex 5.4.

NEW ADVISORY COMMITTEE STRUCTURE

5.29 We have proposed an independent UK expert Animal Feedingstuffs Advisory Committee which we consider should report to Ministers. At various stages of our Report we have identified functions which should be undertaken by this Committee to coordinate controls, surveillance and advice to Government on feedingstuffs issues. These are summarised in Chapter 7.

ENFORCEMENT

5.30 Local authorities are responsible for enforcing the feed marketing and labelling regulations. We found that there were no serious criticisms of their enforcement activities and no serious concerns from local authorities about their competence to fulfil their duties except for powers of entry. We have found their ability to enforce the regulations at various points in the total feed chain was constrained by inadequate powers, usually powers of entry, at ports and on farms. Our recommendations at 5.8, 5.9, and 5.22 are designed to rectify this.

ANNEX 5.1

LEGISLATION: FEED COMPOSITION AND MARKETING

THE AGRICULTURE ACT 1970

5.1.1 The primary legislation governing the controls on feedingstuffs is the Agriculture Act 1970 (Part IV). Its main requirements are that feedingstuffs when sold should be fit for their intended purpose and free from harmful ingredients. A statutory ingredients statement is required as to the composition of feed and information and instructions must be provided on storage, handling and use.

5.1.2 There are two sets of secondary regulations on animal feedingstuffs made under Part IV of the 1970 Act.

THE FEEDING STUFFS REGULATIONS 1991

5.1.3 The Feedingstuffs and the Feedingstuffs (Sampling and Analysis) Regulations 1982 (as amended) [see Chapter 6 for details of these regulations] contain detailed schedules. There is parallel legislation in Northern Ireland. The Regulations for Northern Ireland corresponding to the Feeding Stuffs Regulations 1991 have been made this year. The regulations set:

— definitions of various types of feedingstuffs, feed ingredients, feed additives or other contents of feedingstuffs, storage life and a definition of the national list of feed manufacturers.

— the form of the statutory statement which must appear on the product label or be supplied with feedingstuffs.

— packaging requirements.

— conditions and restrictions on the sale and use of feed additives, protein sources and feedingstuffs, and limits for undesirable substances.

5.1.4 Detailed provisions in the schedules include:

Schedule 1, Part I

— the content of the statutory statement. The label must include items such as a description of the feed, quantity, storage life, the additives and ingredients it contains and moisture content when it exceeds a certain level. Additionally the statutory statement may list information about price, country of origin, directions for use, date of manufacture, type of processing or treatment. Label information must not mislead or make unsubstantiated claims.

Schedule 1, Part II

— lists analytical constituents for which labelling is compulsory and for which labelling is optional, the type of feedingstuffs and species of animal to which they relate.

Schedule 2

— lists permitted names and descriptions of straight feedingstuffs and materials sold for animal feedingstuffs together with certain compulsory and optional labelling information for such sales; the list is not exclusive and unlisted materials can also be sold as feedingstuffs.

Schedule 3

— stipulates limits of variation which can be applied to label declarations of the content in compound and other feedingstuffs of fibre, oil, ash, protein, moisture, certain amino acids, trace elements, vitamins, and energy.

Schedule 4 — contains lists of permitted antioxidants, colourants, emulsifiers, stabilisers, thickeners and gelling agents, binders, anti-caking agents and coagulants, vitamins, pro-vitamins and substances having a similar effect, trace elements, aromatic and appetising substances, preservatives, acidity regulators; for some additives there is a restriction on the maximum content of an additive and the species of animal to which it may be fed.

Schedule 5 — prescribes limits for specific undesirable substances in feedingstuffs. They include heavy metals, pesticides, nitrates, aflatoxin and certain seeds and fruits.

Schedule 6 — lists categories of feed ingredients which must appear on feed labels when the full list of ingredients is not given.

Schedule 7 — some permitted protein sources must meet certain compositional criteria or are restricted for feeding to certain animal species.

Schedule 8 — lays down labelling instructions for additives and premixtures.

Schedule 9 — sets out the methods for calculating the energy value of poultry, ruminant and pig feeds.

ANNEX 5.2

INTERDEPARTMENTAL COMMITTEE ON ANIMAL FEEDINGSTUFFS

The Interdepartmental Committee on Animal Feedingstuffs (IDCAFS) was set up in 1982. Its functions are:—

"to advise on all aspects of animal feedingstuffs; co-ordinate views on policy as they concern feedingstuffs for the benefit of delegates to EC Committee meetings in Brussels; guidance for MAFF officials in the formulation of national controls. It examines information on feedingstuffs ingredients, additives, undesirable substances, bio-proteins and some aspects of labelling requirements for feedingstuffs".

Members

Chairman

Mr M Stranks (until March 1992)	Senior Professional Officer Nutrition Chemistry, Agricultural Development and Advisory Service (ADAS)
Mr A E Buckle	Ministry of Agriculture, Fisheries and Food (MAFF) Central Science Laboratory, Slough
Dr D I Givens	Director of the ADAS Feed Evaluation Unit
Mr A Hacking	ADAS, Welsh Office Agriculture Department
Dr D A Jonas	Food Science Division II, MAFF
Dr C A Lawrie (from January 1992)	Food Science Division I, MAFF
Dr D Lees (until January 1992)	Food Science Division I, MAFF
Mr C T Livesey	Biochemistry Discipline, Central Veterinary Laboratory
Ms J Manson	Health and Safety Executive
Dr H Marquand	Department of the Environment
Mrs A McDonald	Department of Health
Mr A Robb	Scottish Office, Agriculture and Fisheries Department
Mr K M Treves-Brown	Veterinary Assessor in the Veterinary Medicines Directorate
Dr E Unsworth	Department of Agriculture for Northern Ireland
Mr A Wotherspoon	Food Science Division I, MAFF

ANNEX 5.3

STEERING GROUP ON CHEMICAL ASPECTS OF FOOD SURVEILLANCE

The Steering Group on Food Surveillance was established in 1971, and changed its name to The Steering Group on Chemical Aspects of Food Surveillance in 1991 to distinguish it from the newly-founded Steering Group on the Microbiological Safety of Food. It provides advice on the chemical safety and nutritional adequacy of the UK food supply.

Its terms of reference are:

"To identify through surveillance the need for action to ensure the chemical safety and nutritional adequacy of food."

Members

Chairman

Dr W H B Denner	Chief Scientist (Food), Ministry of Agriculture, Fisheries and Food (MAFF)

Deputy Chairman

Dr J R Bell	Food Science Division I, MAFF
Professor D Barltrop	Dept of Child Health, Westminster Hospital, University of London
Professor B W Bycroft	Dept of Pharmaceutical Sciences, University of Nottingham
Mr E C Davison (Alternative Scottish representative)	The Scottish Office Agriculture and Fisheries Dept
Mr B H B Dickinson	Food Safety Group, MAFF
Dr G E Diggle	Health Aspects of Environment and Food Division, Department of Health (DH)
Mr A J Harrison	Public Analyst, County of Avon Scientific Service, Bristol
Professor R M Hicks	United Biscuits (UK) Ltd, Group Research and Development Centre
Professor S R Jones	Head of Environmental Studies, British Nuclear Fuels plc, Sellafield
Dr N J King	Department of Environment, Central Directorate of Environmental Protection
Miss L Lockyer	Health Aspects of Environment and Food Division, DH
Dr A F MacLeod	Scottish Office Home and Health Department
Dr C H McMurray	Chief Scientific Officer, Department of Agriculture for Northern Ireland
Mr P W Murphy	Pesticides, Veterinary Medicines and Emergencies Group, MAFF
Dr P I Stanley	MAFF, ADAS Central Science Laboratory, Slough

65

ANNEX 5.4

SCIENTIFIC COMMITTEE FOR ANIMAL NUTRITION

The Scientific Committee for Animal Nutrition (SCAN) was created by Commission Decision 76/791/EEC of 24 September 1976 in order to provide the EC Commission with expert advice on scientific and technical matters related to animal nutrition and related aspects, as described in Article 2 of the Commission Decision:

"1. The Committee may be consulted by the Commission on scientific and technical questions relating to the nutrition and health of animals and to the quality and wholesomeness of products of animal origin.

In particular, the Committee may be consulted on questions concerning the composition of feedingstuffs, processes which are liable to modify feedingstuffs, additives, and substances and products which may be considered undesirable in feedingstuffs.

2. The Committee may draw the attention of the Commission to any such problem."

Members

Chairman

Professor M Vanbelle	Laboratory of the Biochemistry of Nutrition, Dept of Applied Animal Biology, Catholic University of Louvain

Vice-Chairman

Professor P S Elias	Federal Research Centre for Nutrition, Karlsruhe
Professor A Anadón	Dept of Pharmacology, Faculty of Medicine, University "Complutense" (Alcalá de Henares)
Professor C Beretta	Director of the Institute of Veterinary Pharmacology and Toxicology, Faculty of Veterinary Medicine, University of Milan
Dr G Bories	Director of the Laboratory of Xenobiotics, National Institute of Agronomic Research (INRA), Toulouse
Professor J W Bridges	Director of Robens Institute of Industrial and Environmental Health and Safety, University of Surrey
Professor E Castellà Bertran	INIA Sub-directorate General of Animal Health, Madrid
Professor V Elezogiou	Department of Pharmacology, Veterinary Faculty, University of Thessalonika
Professor T H Fernandes	School of Veterinary Medicine, Lisbon
Professor D Kutter	Laboratory of Clinical Analyses, Luxembourg
Professor B B Nielsen	Ministry of Agriculture, State Veterinary Serological Laboratory, Copenhagen
Dr D M Pugh	Veterinary College of Ireland, Dublin

Professor D Sauvant	National Agronomic Institute, Paris-Grignon, Paris
Professor P C Thomas	Principal and Chief Executive, Scottish Agricultural College, Edinburgh
Professor F Ungemach	Institute for Pharmacology and Toxicology, Faculty of Veterinary Medicine, Free University of Berlin
Professor F Valfré	Faculty of Veterinary Medicine, University of Milan
Dr P W Wester	Pathology Laboratory, Royal Institute for Public Health and Environmental Hygiene, Bilthoven

CHAPTER 6

METHODS OF ANALYSIS

INTRODUCTION

6.1 Methods of analysis have been dealt with sector by sector, corresponding to the legislation examined in the separate Chapters in the report. Areas examined include progress on an ELISA (enzyme-linked immunosorbent assay) test for the enforcement of bovine spongiform encephalopathy (BSE) legislation and the question of the development of rapid methods for the detection of salmonella in feed materials. The absence of comprehensive statutory analytical methods for the enforcement of the legislation on feed composition and marketing and medicated feedingstuffs has also been considered.

BOVINE SPONGIFORM ENCEPHALOPATHY

6.2 There is currently no method of analysis available for detecting ruminant protein in ruminant rations and therefore for the enforcement of the BSE Order 1991 (see paras 2.14 and 2.15). However, an ELISA system has been developed by MAFF. It has proved successful in tests involving meat and bone meal and it has been possible to differentiate between bovine and ovine material in samples processed below temperatures of 130°C. The method has now been validated and can be applied to compound feedingstuffs. **We welcome the development of this test.**

6.3 DNA (deoxyribonucleic acid) probes are also being developed to determine the species from which animal protein has been derived. This test may be restricted in its availability in the immediate future. DNA probes would be sufficiently sensitive to detect the low level of protein in refined tallow derived from ruminants when incorporated into feedingstuffs. This would need to be taken into account if this method is to be used to enforce the ruminant protein ban, a problem unlikely to be presented by the ELISA test.

SALMONELLA

6.4 As outlined in Annex 3.4, sampling under the Processed Animal Protein Order (PAPO) is undertaken by MAFF on twenty days' production over the year ('official' samples), and also by registered processors, who are required to take samples of their products each day, as they are dispatched from the premises ('private' samples). The sampling method is laid down in PAPO. Official samples are tested by MAFF SVS at one laboratory, while private samples must, under PAPO, be analysed by laboratories approved by MAFF. All approved laboratories must employ one of two methods described in Schedule 1 to PAPO — a bacteriological method or one based on electrical conductance. These two methods are also recommended for the testing of samples taken by manufacturers under the five codes of practice for the control of salmonella in feed materials, outlined in Annex 3.4.

6.5 In general, feed materials sampled under the codes of practice for salmonella are dispatched from manufacturers' premises shortly after processing to reduce the need for extensive storage facilities for finished products. This may also occur with processed animal protein at some premises. The bacteriological method takes a number of days to produce an initial result. Although the electrical conductance method is faster, some feed materials will have already entered the feed/food chain by the time the analytical results using this method have become available. Any action arising from the isolation of salmonella, in these circumstances, could only be applied retrospectively, in order to prevent a recurrence of contamination. **We therefore recommend that research into more rapid methods of analysis for salmonella in feed materials be pursued as a high priority to help ensure that contaminated material can be prevented from entering the feed/food chain.**

6.6 At an EC level, reference methods are currently being drawn up for the detection of salmonella (as well as for clostridia and *Enterobacteriaceae*). They will be for use by the authorities enforcing EC Directive 90/667 on the disposal and processing of animal waste (see Para 2.16).

MEDICATED FEEDINGSTUFFS AND FEED COMPOSITION AND MARKETING

6.7 Many of the arrangements concerning the development and application of methods of analysis for the enforcement of these two areas of legislation are common since, in both cases, the methods are used to determine the chemical composition (as opposed to the microbiological status) of the feed. At EC level, they are dealt with by the same analytical methods committee (with the exception of Prescription Only Medicines (POMs) (see Para 6.9)). In the absence of statutory EC methods, other valid methods can be used. In the UK, the Laboratory of the Government Chemist (LGC) is largely responsible for the development and up-dating of methods either to produce national statutory methods, where these apply, or to aid in the UK's negotiations within the EC. It is also designated as the 'referee' analyst in the event of litigation. This function is undertaken in Northern Ireland by the Chief Agricultural Analyst.

Legislation on sampling and analysis of medicated feedingstuffs

6.8 Methods of sampling and analysis for the enforcement of medicated feedingstuffs legislation are laid down in the Medicines (Animal Feedingstuffs) (Enforcement) Regulations 1985. They are used to ensure that the identity and concentration of medicinal additives incorporated into marketed feedingstuffs conforms with the declaration on the label. The Enforcement Regulations cover intermediate as well as complete feedingstuffs, feed supplements and protein concentrates. Sampling in Great Britain is undertaken by local authorities and in Northern Ireland by the Department of Agriculture. The analyses are performed by public analyst laboratories. (A further tier of enforcement activity consists in the sampling and analysis of animal products for veterinary residues (see para 4.19)).

6.9 The differing licensing arrangements for POMs and PMLs (Pharmacy and Merchants List products) (see paras 4.6-4.8) are reflected in the mechanisms used to develop statutory methods for their analysis. While statutory methods for PMLs are dealt with by an EC committee, the UK has discretion to lay down statutory methods for POMs on a national basis.

Sampling and analysis: feed composition and marketing

6.10 The regulations on methods of analysis are contained in the Feeding Stuffs (Sampling and Analysis) Regulations 1982 (as amended). They contain methods for certain additives and feed constituents, such as protein, fibre and ash, for the enforcement of the labelling provisions, and also methods for certain undesirable substances. All these have been agreed on an EC basis. No statutory method of analysis is as yet prescribed for the enforcement of ingredients declaration (see Para 5.4).

Problems associated with EC methods of analysis

6.11 There was general concern expressed in the comments submitted to us, both by trade organisations and enforcement agencies, about the slowness with which new developments were reflected in the analytical regulations. There were complaints that some of the methods in the Regulations required up-dating, or in the case of both medicinal and technological additives, were for substances no longer used commercially. On the other hand, for many widely used additives no statutory methods were available, in some cases even many years after approval of the additives for use. Methods were also required for a range of declarable feed constituents, and for a large proportion of the undesirable substances for which maximum limits are laid down. Progress on the development of statutory EC methods of analysis is often slow, because of the difficulties in coordinating the validation of methods by laboratories across the Community. However, this situation has been made worse because the EC committee responsible for approving methods of analysis ceased to function for a number of years. We understand that this has now been rectified, but we are concerned that this situation could have arisen.

6.12 We examined the difficulties in enforcement presented by the absence of statutory methods of analysis. In many cases, although statutory methods are not available, there are methods recognised by official bodies which are often acceptable in the event of litigation. However, we recognise that there may be some instances where prosecutions are made more difficult by the absence of an official method. We therefore welcome the revival of the EC committee work on methods of analysis.

6.13 We examined the desirability of developing national statutory methods for the enforcement of EC requirements, in the absence of EC methods. We were advised, and accepted, that it might not be possible to do so in a legislative area where the EC has responsibility.

Methods of analysis for POMs

6.14 Although the UK has discretion to lay down statutory methods for POMs on a national basis, it does not exercise this right. As a result, there are few methods of analysis for POMs in the Medicines (Animal Feedingstuffs) (Enforcement) Regulations and those which do exist require up-dating, or are for products no longer used commercially. **As some POMs are used extensively, and the absence of up-to-date statutory methods may provide a barrier to enforcement, we recommend that new statutory methods for POMs be introduced, and that those that exist are up-dated, where necessary.**

6.15 Under Directive 70/524, manufacturers, when presenting a new additive to the EC for approval (whether medicinal or technological), are obliged to present a method of analysis for its identification. This method, although sometimes unsuitable for enforcement purposes in the form presented, is often used as the basis for the development of a statutory method by the EC Analytical Methods Committee. Methods of analysis for additives approved under this directive are explicitly excluded from any provisions regarding commercial confidentiality. In the case of POMs, manufacturers are also obliged to provide a method of analysis for any additive presented for approval. However, the methods for these medicinal additives may only be provided to bodies other than the Veterinary Products Committee (including enforcement authorities) with the permission of the manufacturer. As this may potentially hinder the enforcement of the medicated feed legislation, including the development of any statutory methods, **we recommend that a mechanism be put in place to ensure that enforcement authorities, and the bodies responsible for the development of statutory methods, automatically have access to the method of analysis of any medicinal additive once it has been licensed in the UK.**

Tolerances

6.16 Tolerances (limits of variation) are limits to the amount by which certain feed constituents can vary from the level declared on the feedingstuffs label. They are laid down in the Feeding Stuffs Regulations 1991, and the Medicines (Feeding Stuffs) (Limits of Variation) Order 1976. Variations can arise in the material itself at the manufacturing stage, in sampling, and through methods of analysis, both through variations in results using a given method within the same laboratory, or in different laboratories.

6.17 Individual member states have some scope to set their own tolerance limits for analysis, even for those feed constituents covered by EC provisions, as the EC sets only minimum tolerances. Against this background, we *welcome* the initiative by the LGC, the Analytical Methods Committee of the Royal Society of Chemistry (RSC) and the industry to study the sources of analytical variation in selected medicinal additives. This will go some way to ensure that decisions regarding tolerances are based on scientific data and good manufacturing practice. Additionally, **we recommend that a mechanism be put into place to formalise the procedures whereby national tolerances are set, possibly by seeking advice from the Animal Feedingstuffs Advisory Committee.**

Analytical Committees

6.18 The UK has three non-governmental committees on analytical methods in animal feedingstuffs. The first two are sub-committees of the RSC Analytical Methods Committee, one deals with medicinal additives, and the other with other feed constituents, and undesirable substances. They both organise collaborative studies, assess results and focus attention on areas of difficulty and on particular problems arising in laboratories or the Courts. Membership includes trade organisations, enforcement bodies, the LGC and MAFF. MAFF provides some financial support. The third is under the auspices of the British Standards Institute. It does not cover medicinal additives. It meets periodically to discuss draft standards and to prepare the UK position for International Standards Organisation meetings. This committee structure appears to operate effectively without serious shortcomings.

ANNEX 6.1

LEGISLATION: METHODS OF ANALYSIS

INTRODUCTION

6.1.1 Methods of analysis are laid down in secondary legislation, in parallel with the legislation they are designed to enforce.

BOVINE SPONGIFORM ENCEPHALOPATHY

6.1.2 There are no statutory methods for the enforcement of the BSE Order 1991. However, an ELISA test capable of detecting ruminant protein in feedingstuffs for ruminants has recently been developed.

SALMONELLA

6.1.3 Methods of sampling and analysis for the detection and isolation of salmonella in processed animal protein are laid down in the Processed Animal Protein Order 1989 (PAPO). They are prescribed for use both by agricultural department officials who take samples for testing twenty days a year at the processing plant, and by registered processors, who must sample processed animal protein every day it is despatched from the plant. These samples must, under PAPO, be analysed by laboratories approved by MAFF using one of two methods — a bacteriological method and one based on electrical conductance. These methods are also recommended for the testing of samples taken by manufacturers under the five codes of practice for the control of salmonella in feed materials (see Annex 3.4). Parallel provisions for Northern Ireland are laid down in the Diseases of Animals (Animal Protein) (No 2) Order (Northern Ireland) 1989 (as amended). The Department of Agriculture (DANI) is responsible for enforcement of the legislation, including the taking of official samples, and for the approval of laboratories.

MEDICATED FEEDINGSTUFFS

6.1.4 Methods of sampling and analysis for medicated feedingstuffs are laid down in the Medicines (Animal Feeding Stuffs) (Enforcement) Regulations 1985 which are applicable throughout the UK. Sampling is undertaken by local authorities in Great Britain, and analysis performed by public analysts on their behalf. In Northern Ireland, sampling and analysis is undertaken by DANI. Tolerances are laid down in the Medicines (Animal Feeding Stuffs) (Limits of Variation) Order 1976. The Regulations are not exhaustive: there are many approved medicinal additives for which no method of analysis is laid down. In the absence of a statutory method, other valid methods are used for enforcement purposes.

FEED COMPOSITION AND MARKETING

6.1.5 In Great Britain, the methods for the enforcement of the Feedings Stuffs Regulations 1991 are contained in the Feeding Stuffs (Sampling and Analysis) Regulations 1982 (as amended). The Regulations contain methods of analysis for certain feed components and non-medicinal additives, for the enforcement of labelling provisions, and also methods for certain undesirable substances. Tolerances (limits of variation) are laid down in the Feeding Stuffs Regulations 1991. As with medicated feedingstuffs, responsibility for sampling and analysis rests with local authorities, and, in the absence of statutory methods, other valid methods are used for enforcement purposes. The parallel legislation in Northern Ireland is the Feeding Stuffs (Sampling and Analysis) Regulations (Northern Ireland) 1982 (as amended). Responsibility for sampling and analysis rests with DANI.

CHAPTER 7

CONTROLS AND THE FEED CHAIN

THE FEED CHAIN

7.1 The feed industry comprises importers and processors of raw materials; merchants; suppliers of by-products from agro-industrial operators and food manufacturers; feed additive and supplement manufacturers; commercial compounders; integrated units and on-farm mixers. A background note on the feed industry is at Annex 7.1. The functional structure of the industry is shown in Annex 7.2.

7.2 The preceding Chapters have examined the effectiveness of legislation on Bovine Spongiform Encephalopathy (BSE) and other Spongiform Encephalopathies (SEs), Salmonella and other Pathogens, Medicated Feedingstuffs, Feed Composition and Marketing and Methods of Analysis. This Chapter summarises the effectiveness of controls on the different parts of the feed chain. It also considers the mechanisms by which food safety requirements are taken into account in regulating the industry.

RAW MATERIALS

7.3 The feed industry is a user of numerous by-products of other industries. Some, such as cereals, are used untreated, whilst others such as oil seeds are available as cakes or meals following processing to extract their oil. General legislative controls require that feedingstuffs, including raw materials, when sold, should be fit for their intended purpose, and free from harmful ingredients. Raw materials should also comply with a voluntary code of practice for the control of salmonella.

7.4 Certain raw materials, such as specified offals, faeces and urine are prohibited and ruminant protein has restricted uses in order to protect human and animal health.

7.5 Materials which are particularly vulnerable to microbiological contamination have specific controls. Imported animal protein must be licensed and domestically-produced animal protein must comply with statutory sampling and testing arrangements. Industries which produce animal protein are subject to voluntary codes of practice for the control of salmonella.

7.6 Certain raw materials — maize, cottonseed, copra, groundnut, babassu, palm kernel and their derivatives — are vulnerable to aflatoxin contamination. When the aflatoxin level exceeds 0.05, but is still below 0.2 mg/kg, the raw material must be labelled with the aflatoxin content and sold only to approved buyers who have mixing and storage facilities which can guarantee their ability to mix a final feed which meets the statutory limits of 0.05 mg/kg or lower.

7.7 Bioproteins and products of biotechnology are evaluated for safety and efficacy by expert committees before they are permitted for use in feedingstuffs. There are no requirements for other feed materials to be similarly assessed.

7.8 None of the provisions enacted under the Agriculture Act apply at ports. Over 3 million tonnes of raw materials are imported annually for use in feedingstuffs. Port Health Authorities who traditionally enforce the provisions of human food legislation at ports are not empowered to enforce legislation on raw materials intended for feedingstuffs (a small quantity of finished feedingstuffs is imported). Trading standards officers who are designated enforcement officers of feedingstuffs legislation made under the Agriculture Act are not generally empowered to operate at ports. We consider that powers are necessary to sample imported materials and we have recommended this in Chapter 5.

7.9 We examined the effect of the general requirements referred to in paragraph 7.3 on trade in raw materials. We were told that commercial contracts usually offer guarantees of protein and oil content rather than end-use suitability. Additionally contracts might occasionally require consignments to be salmonella-free. A new draft EC proposal is designed to strengthen controls on trade in raw materials. It will introduce an early warning system about movements of contaminated consignments. Other measures being considered by the EC Commission would require raw material suppliers to give information about the degree of contamination present. We think this should benefit all parts of the feed chain.

7.10 Imported raw materials are one source of new strains of salmonella for our animal and human population. Vegetable protein has recently been found to be a source of salmonella infection through surveys at ports and at compounders premises. We expect these findings to have alerted feed manufacturers to the potential risk from vegetable materials but we think the situation should be watched and we have recommended further monitoring. Some current investigations by the Central Veterinary Laboratory (CVL) have confirmed that there may be problems at rapeseed crushing plants. These are being investigated. We have not made recommendations for legislative controls on vegetable materials in anticipation of further survey results and action by the EC Commission to harmonise controls for salmonella specifically in feedingstuffs containing vegetable material.

7.11 The current restrictions on feed ingredients have been developed in response to particular concerns, notably BSE. **We think that there should be a continuous monitoring of trends in raw materials usage and, in particular, an examination of own-species recycling to assess whether there are unforeseen risks of disease, other than SEs, which are being separately examined by the Spongiform Encephalopathy Advisory Committee. We recommend that the new Animal Feedingstuffs Advisory Committee should undertake these functions.**

COMMERCIALLY MANUFACTURED FEEDINGSTUFFS

7.12 Feedingstuffs legislation distinguishes between a feedingstuff which is intended for sale and a feedingstuff which is not. The more extensive controls apply, uniformly, to all feedingstuffs intended for sale. They are described fully in the various annexes to the main chapters of this Report. In addition to statutory controls there are also voluntary codes of practice for salmonella for feed manufacturers to follow. They may also be required, by farmers or retailers, to comply with various product quality assurance schemes or to have checks or monitoring in place which might satisfy the due diligence provisions of the Food Safety Act 1990.

7.13 We noted the large feed manufacturers apply a degree of self-regulation. They purchase raw materials selectively and check the finished products. Several companies are introducing quality assurance schemes to acquire accreditation under BS 5750. These efforts and the provisions of BS 5750 cannot in our view be reasonably matched by smaller or less well- equipped manufacturers. We have not recommended any change to the controls as they currently apply to feed manufacturers who sell feed. However we support the principle of good manufacturing practice and believe that such schemes can with benefit be suitably adapted for smaller feedingstuffs manufacturers. We consider that guidelines on both good manufacturing practice and hazard analysis principles can best be developed in codes of practice. We have recommended this.

HOME-MIXED FEED

7.14 Feedingstuffs manufactured within integrated units and on farms is not usually sold. The 40% of the total feed supply manufactured in this way must comply with statutory limits for pesticides, heavy metals and aflatoxin. It must also exclude ruminant protein in ruminant feedingstuffs and specified offal and ingredients such as faeces and leather etc.

Manufacturers who incorporate medicinal additives must also, like the commercial compounders, register with the Royal Pharmaceutical Society of Great Britain (RPSGB) and the Department of Agriculture for Northern Ireland and comply with a statutory Code of Practice. There are currently about 4,600 small manufacturers on the registers. Some feed manufacturers are registered with MAFF in order to use certain contaminated raw materials. They should also be subject to inspection.

7.15 We think that manufacturers who mix feed for their own use should fulfil their statutory obligations but there are gaps in their knowledge of these controls. These shortcomings can be overcome by including the appropriate information in codes of practice which also outline the principles of good manufacturing practice and hazard analysis as we have recommended in paragraph 5.10. This feed should be checked to ensure it complies with the regulations and we have recommended that legal powers should be adequate to facilitate this.

FEEDINGSTUFFS COMMITTEES

Advisory Committees

7.16 There are two Committees, the Veterinary Products Committee (VPC) and the Interdepartmental Committee on Animal Feedingstuffs (IDCAFS) which have a responsibility to advise on feedingstuffs matters. The VPC is an independent scientific committee which deals solely with veterinary medicinal issues (see paragraph 4.3). IDCAFS is a committee of officials which advises on technological additives and on some general feedingstuffs issues, particularly those relating to policy on feed composition and marketing. The Spongiform Encephalopathy Advisory Committee, may, when necessary, give advice on feedingstuffs. Permanent committees, the Food Advisory Committee (FAC), the Advisory Committee on Pesticides (ACP) and the Committee on Toxicity of Chemicals in Food, Consumer Products and the Environment may also on occasions give advice which has implications for feedingstuffs.

Feed Surveillance Committees

7.17 There are independent committees which coordinate food surveillance. The Steering Group on the Chemical Aspects of Food Surveillance identifies survey needs to assess chemical contamination of food and the Steering Group on the Microbiological Safety of Food has similar responsibilities for microbiological contamination of food. We think that feedingstuffs should be regarded as part of the food chain for surveillance purposes. We have made recommendations which would incorporate feedingstuffs into the existing arrangements for food.

New Animal Feedingstuffs Advisory Committee

7.18 **No single committee has an overview of all feedingstuffs issues and we recommend that an Animal Feedingstuffs Advisory Committee be established.** There is a need for a single committee to be aware of, and consider, all technical developments in raw materials, manufacturing processes and feed treatments. It should coordinate and advise on policy and legislative measures and on EC proposals. It should also keep under review and advise on the nutritional characteristics and the safety of feed ingredients; the safety, quality, and efficacy of feed additives; undesirable substances in feedingstuffs and recycling of own-species material. The Animal Feedingstuffs Advisory Committee should have links across to the Veterinary Products Committee, the Spongiform Encephalopathy Advisory Committee, the Advisory Committee on Pesticides so as to be informed of any action they may take which affects animal feedingstuffs. It should also encompass the functions currently undertaken by IDCAFS. The new Committee will also need to be aware of products and processes with probable or potential use in feed and particularly those which have been considered by the Advisory Committee on Releases to the Environment. The Animal Feedingstuffs Advisory Committee should keep a watching brief on changes in feed manufacturing practices which might have implications for the work of the Spongiform Encephalopathy Advisory Committee and the Veterinary Products Committee. It should also see and act upon the reports of surveys recommended or coordinated by the Steering Group on the Microbiological Safety of Food and the Steering Group on Chemical Aspects of Food Surveillance. It should also be able to recommend to these Committees areas which require attention. Members of the Animal Feedingstuffs Advisory Committee should have appropriate expertise in human and animal health and agricultural practice. The Animal Feedingstuffs Advisory Committee should be an independent committee; it should report to Ministers and publish reports.

ANNEX 7.1

THE FEED INDUSTRY

LIVESTOCK PRODUCERS' FEED COSTS

7.1.1 Livestock producers buy in £7.0 billion of farm requisites. There are about 120,000 livestock producers in the UK and feed represents the largest cost (at £2.67 billion, 41% of the total). The MAFF Publication 'Agriculture in the UK' (previously the Annual Review White Paper) of 1991 quoted expenditure of £2.07 billion on compound feeds and £724 million on straight feeds.

STRUCTURE OF THE FEED INDUSTRY

7.1.2 The functional structure of the feed industry is shown in Annex 7.2. About 55% of all purchased feeds are cereals (about 40% of which are compounded) and they mainly come from UK grain merchants, traders and farms. Most of the other 45% of feeds and feed ingredients are by-products from agro-industrial operators (millers, maize processors, distillers etc) divided between UK and imported supplies. These materials flow to the farms through a variety of channels. Two thirds (by volume) flow through the compounders, and one third through agricultural merchants, blenders, mobile mixers or direct to farmers from retailers of straights. Most compounds are sold direct to farmers (70% of production) and 30% (mainly from the two largest compounders) is distributed through the agricultural merchant trade. An increasing proportion of livestock is fed on rations mixed on farm using straight feedingstuffs. Cattle and pig rations contain home-grown cereals to which are added purchased proteins, additives, supplements and oils and fats. Large integrated poultry units normally buy in all ingredients and mix feed themselves. In 1987, purchased concentrates (non compounds) including straights totalled 4.5m tonnes for all livestock. This compared with 10.6m tonnes production from compound mills.

Feed production

7.1.3 The average annual production of compound feed has remained fairly constant for the last 20 years but the product mix has changed considerably. In 1973, 59% of compound feed was accounted for by pig and poultry feed production with cattle feed accounting for a further 35%. Sixteen years later the proportion of compound feed accounted for by pig and poultry feed had fallen to 54% while that for cattle feed had risen to 39%. A number of factors influenced this shift including the CAP price support system for milk and beef, the absence of such support for pigs and poultry together with the declining market for eggs and consequently layer feeds.

Feed compounders

7.1.4 The structure of the feed industry comprises three categories of compounders. The first category contains the three companies who own mills throughout the country and who are consequently called 'nationals'. The three companies are Dalgety Agriculture (Dalgety plc), Bibby (Barlow Rand SA), and Pauls Agriculture/ BOCM (Harrison Crosfield). These companies have vertical links (usually through parent companies) either forward to livestock production and marketing or backwards to raw material sources (or both). Together these three companies commanded between 50% and 55% of the market in the late 1980s. The second category comprises independent or 'country' compounders. They can be regional compounders or be single family-owned mills. The third category is the farmer cooperatives which supply about 10% of the market.

Straight feedingstuffs

7.1.5 Non compound feed originates with importers and agro-industrial companies. It is often sold to livestock producers by agricultural merchants and cooperatives. Some larger merchants are also wholesalers. Import of the materials is dominated by 5 multi-national shippers, a further 10 companies ship direct or transship through continental ports and are also involved in wholesale and retailing. It is estimated that 40% of imported materials is transshipped and 60% imported direct compared with figures of 80% and 20% respectively 10 years ago. UK-produced feed materials are processed by multi-national companies in companies employed in maize processing, oilmilling, brewing and distilling and wheat milling industries.

On-farm mixers

7.1.6 There is no reliable source of information about the quantity of materials mixed on farms. MAFF Statistics Division has undertaken various on-farm surveys which have yielded some information. It has, so far, proved impossible to calculate accurately the size of the home-mixer market. Apart from the compound feed total, Annex 7.3 is at best a guide based on estimates by importers and traders in raw materials, and it cannot be substantiated by survey findings. It simply demonstrates that appreciable quantities of cereals (including wheat, barley, and small amounts of rye) move in inter-farm sales and that farmers also buy in ingredients or straights — rapeseed meal, molassed feed, oil cakes etc. It is known that merchants, and more recently some compounders, sell straights to farmers but the amount cannot be quantified at present.

7.1.7 The home mixer typically uses a range of raw materials. The main one is cereals (from UK) but others include rapeseed (from EC and UK), palm kernel meal, soyameal, fishmeal, cottonseed, citrus pulp, sunflower seed (imported), maize gluten (from UK), and a number of cereal by-products such as wheat feed, malt culms, brewers grains, grain screenings in addition to molassed sugar beet.

TRENDS

7.1.8 The animal feed industry has been contracting for some years. The effect of milk quotas in 1984 had a significant effect on cattle feed production and there is a continuing process of rationalisation. Coupled with this rationalisation is the increasing amount of home mixing by livestock producers and by integrated mill and mixing operations.

ANNEX 7.2

FUNCTIONAL STRUCTURE OF ANIMAL FEED INDUSTRY

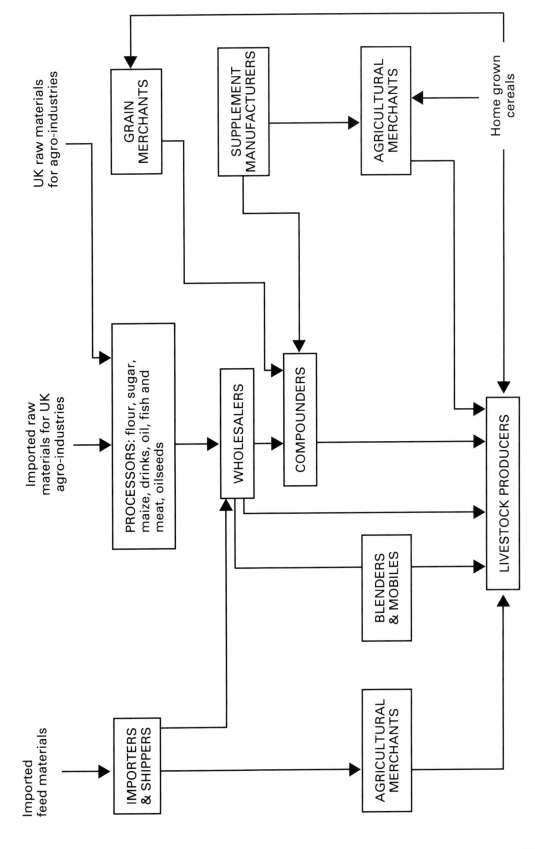

APPROXIMATE FARM USE OF FEED AND FEED MATERIALS 1991

(Fresh weight basis unless specified)

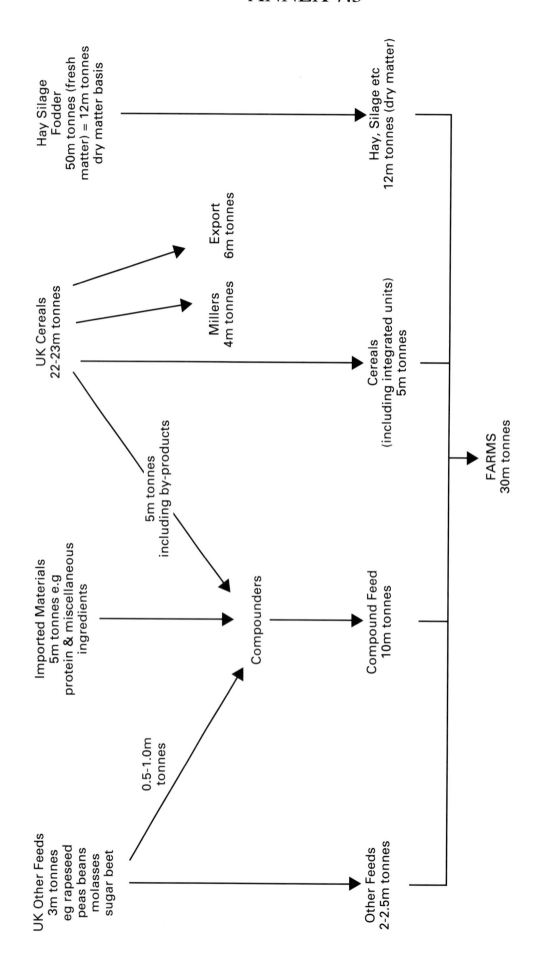

CHAPTER 8

SUMMARY OF RECOMMENDATIONS

INTRODUCTION AND MAIN FINDINGS

8.1 We regard feedingstuffs as an integral part of the feed/farm animal/human food chain for the purposes of ensuring food safety and some of our comments and recommendations contained in the report reflect that view. (paragraph 1.16)

8.2 We found that the feed industry is regulated by extensive controls, which taken together, are in most areas effective in safeguarding food for human consumption. However there are some gaps in legislation and in its enforcement. (paragraph 1.17)

8.3 We recommend that an Independent Animal Feedingstuffs Advisory Committee be established to take an overview of all feedingstuffs issues in relation to human and animal health. We make recommendations concerning remit and scope of work; membership; formal links with other committees, Ministry Departments and surveillance authorities, and method of reporting. (paragraph 1.18 and 7.18)

BOVINE SPONGIFORM ENCEPHALOPATHY AND OTHER SPONGIFORM ENCEPHALOPATHIES

Legislation

8.4 We recognise the importance of the feed bans [on feeding ruminant protein to ruminants and on inclusion of specified bovine offal in animal rations] **in the control of bovine spongiform encephalopathy, and in minimising the exposure of other species and we recommend that they be maintained.** (paragraph 2.13 and 2.16)

8.5 We recommend that the feed bans be retained even after the results of the inactivation study become available, unless the results provide unequivocal information on the inactivation of the scrapie/bovine spongiform encephalopathy agents, and that the necessary conditions can be consistently achieved by the rendering industry.** (paragraph 2.18 and 2.16)

8.6 We recommend that the UK Government seeks to retain national veterinary legislation with reference to EC Directive 90/667 on the disposal of animal waste.** (paragraph 2.16)

Surveillance

8.7 We recommend that surveillance for spongiform encephalopathies be heightened, in order to identify rapidly the development of disease in other species, should it occur.** (paragraph 2.33)

8.8 We recommend that the incidence of bovine spongiform encephalopathy in cattle born after the ruminant protein ban should be carefully monitored.** (paragraph 2.15).

Research

8.9 We recommend that research into the early detection of spongiform encephalopathies in asymptomatic animals continues at the highest level of priority.** (paragraph 2.34)

8.10 We recommend that experimental work on the transmission of bovine spongiform encephalopathy to chickens continues, and the situation reappraised in the light of the results.** (paragraph 2.30)

8.11 We recommend as a matter of priority, that consideration be given to extending the pig oral challenge experiments to include scrapie-infected material.** (paragraph 2.35)

8.12 We recommend that Government and appropriate Committees consider support for research on the species barrier as it relates to spongiform encephalopathies and specifically on the nature of the <u>PrP</u> gene in the pig. (paragraph 2.35)

8.13 We recommend that when the results of the research on the species barrier as it relates to spongiform encephalopathies and the information as a result of notification of scrapie is available, Governments should invite the appropriate committees to examine methods of reducing the reservoir of scrapie. (paragraph 2.36)

Additional Comments	**8.14** We welcome the development of tests for the detection of ruminant protein in meat and bone meal and compound feedingstuffs. (paragraph 2.15)
SALMONELLA AND OTHER PATHOGENS Legislation	**8.15** We recommend that the current controls [on home produced and imported animal protein] be retained and we support the continuing efforts to reduce salmonella contamination. (paragraph 3.32)
Codes of practice	**8.16** We recommend that straight feedingstuffs are subject to a code of practice similar to the existing one for raw materials. (paragraph 3.23)
Information on salmonella	**8.17** We recommend that every effort is made to provide data [on different parts of the feed, animal, human food and human infection chain] which is obtained using methods which are as similar as possible. (paragraph 3.41)

8.18 We recommend that in addition to the existing publications consideration be given to publishing all data on salmonella infection/contamination in a single publication. (paragraph 3.41)

8.19 We recommend that laboratories report positive and negative salmonella results for feedingstuffs and the source of the feedingstuffs to MAFF, and that these results be evaluated. (paragraph 3.31)

Surveillance	**8.20** We regard feedingstuffs as an integral part of the feed/farm animal/human food chain for the purposes of ensuring food safety and consider that the existing arrangements for ensuring the microbiological safety of human food are such that animal feed could, at least in part, be incorporated into the existing arrangements. We specifically recommend that further surveys be carried out to monitor feedingstuffs and vegetable materials and we have identified the Steering Group on the Microbiological Safety of Food as the appropriate body to undertake this work. (paragraph 3.33)

8.21 We also recommend that a survey should be carried out to determine the microbiological quality of feed produced on farms. (paragraph 3.30)

Matters for the Animal Feedingstuffs Advisory Committee	**8.22** We recommend that the results of investigations on levels of salmonella contamination in feed materials at compounders, crushing plants and farms, be examined by the Animal Feedingstuffs Advisory Committee. (paragraph 3.34)

8.23 We recommend that the Animal Feedingstuffs Advisory Committee assesses the use of acid treatment in raw materials and feedingstuffs to control salmonella infection. (paragraph 3.40)

8.24 We recommend that the feasibility of heat treating all poultry feed should be examined by the Animal Feedingstuffs Advisory Committee. Priority should be given to the possibility of heat-treating feed supplies destined for elite and grandparent flocks. (paragraph 3.38)

8.25 We recommend that the Animal Feedingstuffs Advisory Committee should consider all available information including the results of surveys and research on salmonella and other pathogens in feedingstuffs. It should examine the possibility that serotypes found in feed may subsequently become important for animals and man and advise on any measures which need to be taken. Further we recommend that authorities charged with monitoring salmonella and other infection/contamination in the chain comprising animal feed/farm animal/human food and humans be represented on this Committee. (paragraph 3.42)

MEDICATED FEEDINGSTUFFS
Legislation

8.26 We recommend that fish farmers who use medicinal additives should be registered with the Royal Pharmaceutical Society of Great Britain or in Northern Ireland with the Department of Agriculture. (paragraph 4.15)

8.27 We recommend that on-farm mixers using medicinal additives and intermediate medicated feedingstuffs in any manner should be registered with the Royal Pharmaceutical Society of Great Britain or Department of Agriculture for Northern Ireland. (paragraph 4.16)

8.28 We recommend that it be made clear in the regulations and in guidance to the industry that veterinary medicines can only be used in feedingstuffs in accordance with the product licence. (paragraph 4.14)

Licensing of medicinal products

8.29 We recommend that the Veterinary Products Committee agenda and proceedings are made more freely available. (paragraph 4.3)

8.30 We recommend that criteria for Pharmacy and Merchants Lists products and Prescription Only Medicine classification be published as part of the reports of Veterinary Products Committee activities. (paragraph 4.4)

8.31 We note that despite considerable effort by all concerned that sulphonamide residues [in pig kidneys] persist and we recommend that the licensing of drugs for in-feed use which require long withdrawal periods should be carefully considered. (paragraph 4.20)

8.32 We recommend that monitoring of antibiotic resistance of salmonella be continued and that routine monitoring of antibiotic resistance in human *E. coli* be established. (paragraph 4.27)

8.33 We recommend that not only should antibiotics giving cross resistance to those used in human medicine not be used as growth promoters but that their prophylactic use in animals be reconsidered. (paragraph 4.27)

8.34 We recommend that efforts continue to be made by the UK to secure the advice of independent experts in the structure of any European system for approval of medicinal products. (paragraph 4.7)

Additional comments

8.35 We have noted that in future a Member State may not authorise a new active ingredient for use in a veterinary medicine intended for food production animals until a Community-wide maximum residue limit has been established or that it has been determined that a maximum residue limit is unnecessary. (paragraph 4.6)

8.36 We reaffirm that legislative developments [the recent imposition of statutory withdrawal periods for medicinal additives, and statutory maximum residue limits for veterinary residues in food] **do not remove the need for the registration and inspection**

of farm and compounders premises where medicinal additives and intermediate medicated feed are used. (paragraph 4.24)

8.37 We considered that the situation whereby no case can be brought under medicated feedingstuffs legislation against a compounder who fails to incorporate a declared medicinal additive does not pose a risk to human or animal health, and can be satisfactorily dealt with under other 'consumer' legislation. (paragraph 4.28)

8.38 We accept the findings of the Food Advisory Committee on the Report of the Working Party on Veterinary Residues. It was concluded that in general, foods of animal origin contain little or no residues of veterinary products and that where present, they pose no significant risk to human health. It recommended that surveillance be continued and increased in certain areas. We have, however, expressed a reservation concerning sulphonamide residues. (paragraph 4.23)

**FEED COMPOSITION &
MARKETING
Legislation**

8.39 We recommend that powers should be available for the routine inspection of feedingstuffs and raw materials imported from non-EEC countries into the Community via the UK. (paragraph 5.22)

8.40 We recommend that on-farm mixers and integrated units are identified and listed for inspection and that adequate powers are provided so that the feed they produce can be sampled to ensure compliance with statutory requirements. (paragraph 5.8)

8.41 We recommend that there be powers to check that farm businesses registered with MAFF have the proper facilities to deal with contaminated feed materials. (paragraph 5.9)

8.42 We recommend powers be obtained which require manufacturers to notify Government of new feed additives. (paragraph 5.14)

8.43 We endorse measures being promulgated to limit feed contamination and reduce potential risks and we recommend that arrangements are made to ensure that all manufacturers who sell or supply feeds should have full records which will allow feed purchasers to be identified. (paragraph 5.21)

8.44 We recommend rationalisation of marketing and labelling requirements for straights and raw materials. (paragraph 5.6)

Codes of practice

8.45 We recommend that Government provides a code on the principles of hazard analysis and critical control points and good manufacturing practice explaining the responsibilities all feed manufacturers have under the Feeding Stuffs Regulations 1991. (paragraph 5.10)

Surveillance

8.46 We regard animal feedingstuffs as an integral part of the feed/farm animal/human food chain for the purposes of ensuring food safety and consider that the existing arrangements for ensuring the chemical safety of food are such that animal feed could be incorporated into the existing surveillance arrangements under the Steering Group on Chemical Aspects of Food Surveillance. (paragraph 5.24)

8.47 We recommend that the list of banned ingredients [contained in the Feeding Stuffs Regulations 1991], **should be kept under review by the Animal Feedingstuffs Advisory Committee.** (paragraph 5.19)

8.48 We recommend that results of all chemical surveys of feedingstuffs be reported to the Animal Feedingstuffs Advisory Committee. (paragraph 5.25)

8.49 We recommend that, in future, the Animal Feedingstuffs Advisory Committee should assess feed developments and keep under review data on feed composition and nutritional needs of livestock so as to maintain the safety of the food chain. (paragraph 5.12)

8.50 It is recommended that if there is delay in the EC consideration [of certain classes of additives yet to be approved and currently outside EC arrangements] **UK experts should begin to evaluate the products on the market in the UK.** (paragraph 5.14)

8.51 We also recommend that any new class of additives, not covered by EC controls, which appear on the UK market should be considered by the Animal Feedingstuffs Advisory Committee. (paragraph 5.14)

Additional comments

8.52 The introduction of statutory maximum residue limits for pesticides in crops intended for use in feedingstuffs, together with an EC proposal requiring consignments of raw materials to be clearly marked when contamination exceeds specified levels, increases the degree of protection afforded to commercial compounders and on-farm mixers and are to be welcomed. (paragraph 5.20)

8.53 There is clearly scope for differences in view between UK and EC committees which may lead to a different degree of priority being given to the restrictions on the use of particular feed additives. However, we do not believe that this represents a major area of concern or one which requires the introduction of powers beyond those presently available to the UK government. (paragraph 5.18)

METHODS OF ANALYSIS
Development of methods

8.54 We recommend that research into more rapid methods of analysis for salmonella in feed materials be pursued as a high priority, to help ensure that contaminated material can be prevented from entering the feed/food chain. (paragraph 6.5)

8.55 We recommend that new statutory methods for Prescription Only Medicines be introduced, and that those that exist are up-dated, where necessary. (paragraph 6.14)

8.56 We recommend a mechanism to ensure that enforcement authorities, and the bodies responsible for the development of statutory methods have access to the method of analysis of any medicinal additive once it has been licensed in the UK. (paragraph 6.15)

8.57 We recommend that the Animal Feedingstuffs Advisory Committee consider procedures by which national tolerances for feed constituents are set.(paragraph 6.17)

Additional comments

8.58 We welcome the development of the ELISA test for animal protein. (paragraph 6.2)

8.59 The EC committee responsible for approving methods of analysis ceased to function for a number of years. We welcome its revival. (paragraph 6.12)

8.60 We welcome the initiative by the Laboratory of the Government Chemist, the Analytical Methods Committee of the Royal Society of Chemistry and the industry to study the sources of analytical variation in selected medicinal additives. (paragraph 6.17)

CONTROLS AND THE
FEED CHAIN

8.61 We think that there should be a continuous monitoring of trends in raw materials usage and, in particular, an examination of own-species recycling to assess whether there are any unforeseen risks of disease, other than spongiform encephalopathies, which are being separately examined by the Spongiform Encephalopathy Advisory Committee. We recommend that the new Animal Feedingstuffs Advisory Committee should undertake these functions. (paragraph 7.11)

APPENDIX I

EUROPEAN COMMUNITY LEGISLATION

Types of community legislation

1. The Council and the Commission may make regulations; issue directives; take decisions; make recommendations; or deliver opinions.

— Regulations have general application and the direct force of law in all member states. If there is a conflict between a Regulation and existing national law, the Regulation prevails.

— Directives are binding on member states as to the result to be achieved but leave the method of implementation to national governments. They must accordingly be transposed into national law. In the United Kingdom, this may take the form of primary legislation, statutory instruments made under relevant specific powers, or an Order under Section 2(2) of the European Communities Act 1982. In itself, a Directive does not have legal force in the member states, but the European Court of Justice has ruled that where a Directive creates specific rights for individuals, their enjoyment of these rights may not be frustrated by a member state's failure to transpose the Directive by the due date.

— Decisions are binding in their entirety on those to whom they are addressed, whether member states, companies or individuals.

— Recommendations and Opinions have no binding force but state Community views.

In addition the Community institutions may adopt Resolutions. These are non-binding and are designed to give an informal indication of the views of the institution.

2. In general, legislative power in the Communities rests with the Council. But the Treaty gives the Commission the power to legislate in certain circumstances. In addition it is common practice and indeed enjoined by Article 145 of the Treaty, for Council legislation to empower the Commission to take measures to implement the legislation. In such cases, besides defining the scope and nature of the measures, the Council normally imposes procedural requirements on the Commission involving a Committee of officials of member states. The procedures applying to such Committees, generally known as 'comitology', are laid down in a Council Decision of 1987.

The process of legislation

3. Under the EEC and EURATOM Treaties, the Council may, in the great majority of cases, adopt legislation only on the basis of a formal proposal from the Commission. But it is open to the Council to request the Commission to make a proposal. The Commission may amend its own proposal at any time up to the Council's final adoption of the measure.

4. The process of legislation starts with the formal adoption of a proposal by the Commission. In certain cases the Parliament and/or the Economic and Social Committee (an advisory body established under the EEC and EURATOM Treaties) are consulted. The Council is under no legal obligation to take account of the view of either. But, where the Treaty requires the Parliament to be consulted, the Council may not finally adopt a measure until it has received the Parliament's opinion or exhausted the possibilities of obtaining it. In other areas the Cooperation Procedure may apply, giving the Parliament a stronger influence over the content of the legislation. Certain proposals with major financial implications can be the subject to the Conciliation procedure.

5. When a proposal is first tabled it is customary to refer it to a Council Working Group (chaired by the country holding the Presidency) for detailed consideration. In the light of the discussion the Council may adopt the Commission proposals as drafted; request the Commission to amend them; amend them itself by unanimous agreement; reject them, or simply take no decision. The nature of negotiation in the Council will normally vary according to whether the measure concerned can be adopted by qualified majority vote (QMV) or requires unanimity. In the case of unanimity it is immaterial whether the Commission alters its proposal to a form acceptable to the Council or whether the Council acts by amending the proposal. In the case of QMV a measure can be agreed by majority vote if it has the Commission's support; but unanimity is required to amend the proposal against the Commission's wishes. Once proposals have been adopted by the Council, they go to specialist meetings of Jurists/Linguists for final clearance of the text in the official languages of the Community and are then published in the Official Journal.

Parliamentary scrutiny of Community documents

6. All proposals for Community legislation must be deposited in Parliament within 48 hours of receipt in London. The responsible department must provide an explanatory memorandum in a standard form. The proposals are considered in the House of Commons by the Select Committee on European Legislation and in the House of Lords by the Select Committee on the European Communities (commonly referred to as the Scrutiny Committees). Ministers are directly answerable to Parliament for the procedure.

APPENDIX II

GLOSSARY

ADDITIVE
Any substance, or preparation containing any substance, other than a premixture as defined, which, when incorporated into a feedingstuff, is likely to affect its characteristics or livestock production.

AFLATOXIN
Toxin produced by the mould *Aspergillus flavus*.

AGENT
Term used to describe the infectious particles responsible for the spongiform encephalopathies: not yet clearly identified or isolated.

ANIMAL PROTEIN
Defined in the Processed Animal Protein Order (see Annex 3.4).

ANIMAL WASTE
Carcase or parts of animals or fish, or products of animal origin not intended for direct human consumption.

ANTHELMINTICS
Substances administered to expel helminth parasites.

ANTIBIOTICS
Chemical compounds derived from living organisms which are capable in small concentration of inhibiting the life process of microorganisms.

ANTIBIOTIC RESISTANCE
The ability of bacteria to survive in the presence of antibiotics.

ANTIMICROBIAL
A substance with the capacity to inhibit the growth of or to destroy bacteria or other microorganisms whether *in vitro* or *in vivo*.

BIOPROTEIN
A proteinaceous feed ingredient produced by certain technical processes such as fermentation.

BREEDERS
Birds producing eggs intended for hatching, for the purpose of either egg or broiler production ('layer breeders' or 'broiler breeders').

BROILER FLOCKS
Groups of young chickens reared for human consumption as poultry meat (see Annex 3.3).

CLOSTRIDIA
Gram positive spore forming anaerobic rod-shaped bacteria.

COCCIDIOSTAT

A chemical agent added to animal feedingstuffs to retard the life-cycle or reduce the population of a pathogenic coccidium (a type of parasite) to the point where disease is minimised.

EC COMMISSION

Body responsible for the administration of the European Community. It ensures observance of EC rules, has the sole power to propose legislation based on the EC treaties, and implements decisions taken by the Council of Ministers within limits laid down by the Council (further details on its role in the legislative process laid down in Appendix I).

COMPLEMENTARY FEEDINGSTUFFS

A mixture of feedingstuffs which has a high content of certain substances and which, by reason of its composition, is sufficient for a daily ration only if it is used in combination with other feedingstuffs (as defined in the Feeding Stuffs Regulations 1991).

COMPLETE FEEDINGSTUFFS

A compound feedingstuff which, by reason of its composition, is sufficient to ensure a daily ration (as defined in the Feeding Stuffs Regulations 1991).

COMPOUND FEEDINGSTUFFS

A mixture of products of vegetable or animal origin in their natural state, fresh or preserved, or products derived from the industrial processing thereof, or organic or inorganic substances, whether or not containing additives, for oral feeding in the form of complete feedingstuffs or complementary feedingstuffs (as defined in the Feeding Stuffs Regulations 1991).

DIETETIC FEEDINGSTUFF

A feedingstuff for animals with particular nutritional needs.

EC DIRECTIVE

See Appendix I.

EC REGULATION

See Appendix I.

ELITE FLOCKS

See Annex 3.3.

ENDEMIC

(Disease) habitually prevalent in a population.

ENTEROBACTERIACEAE

A large group of bacteria which inhabit the human and animal intestinal tract.

ENTERIC PATHOGEN

A pathogen which causes intestinal disease.

EPIDEMIOLOGY

The study of disease, and disease attributes, in defined populations.

EXOTIC BOVIDAE

Non-domestic members of the family *Bovidae* which includes cows, goats, and sheep. Exotic members include the nyala, greater kudu, eland, gemsbok and Arabian oryx.

FEED COMPOUNDING

The manufacture of finished animal diets from a variety of raw materials.

FEED SUPPLEMENT

A feedingstuff containing a mixture of additives, whether medicinal or nutritional (eg vitamins), for use as an ingredient in compound feedingstuffs, in which it generally constitutes 5% or less of the total quantity.

FINAL MEDICATED FEEDINGSTUFFS

Any substance, not being a medicinal product, which is for use wholly or mainly by being fed to one or more animals for a medicinal purpose, or for purposes that include that purpose, without further processing (as defined in the Medicines (Medicated Animal Feeding Stuffs) Regulations 1992.

GRANDPARENT FLOCKS

See Annex 3.3.

GREAVES

See Annex 2.4.

GROWTH PROMOTERS

Substances which, when given in animal feed, increase feed conversion efficiency or result in better daily liveweight gain, or both.

HAZARD ANALYSIS CRITICAL CONTROL POINT (HACCP)

A systematic approach to the identification of those points in a production operation which may cause harm to the consumer. Those points critical to the safety of the consumer (CCPs) are then monitored and remedial action taken if any deviate from their safe predetermined limits.

HORIZONTAL TRANSMISSION

Spread of infection from one animal to another, through direct or indirect contact with the infected animal.

HISTOLOGY

The microscopic structure of the tissues of animals (or plants).

INTEGRATED UNITS

Livestock units (largely poultry) in which all stages (including the production of animal feedingstuffs), are controlled by the same company.

INTERMEDIATE MEDICATED FEEDINGSTUFF

A medicated feedingstuff sold, supplied or imported for use wholly or mainly as an ingredient in the preparation of a substance which is to be fed to one or more animals for a medicinal purpose or for purposes that include that purpose, with or without further processing (as defined in the Medicines (Medicated Animal Feeding Stuffs) Regulations 1992).

ISOLATE or ISOLATION

A single species of microorganism originating from a particular sample or environment growing in pure culture.

KNACKERS
Processors of animal carcases, particularly fallen or casualty stock, which are often processed into pet food. The hides are sold and the offals usually sent for rendering.

LAYER FLOCKS
Groups of domestic fowl kept for egg production.

MATERNAL TRANSMISSION
The spread of infection from mother to offspring *in utero*, or immediately after birth.

MEAT AND BONE MEAL
See Annex 2.4.

MULTIPLE ANTIBIOTIC RESISTANCE
Resistance of bacteria to antibiotics of different chemical groups.

ON-FARM MIXERS
Farmers who manufacture finished feedingstuffs for their own use rather than buying them from a commercial compounder.

PARENT FLOCKS
See Annex 3.3.

PARENTERAL
By a route (eg of infection) other than oral.

PASTEURISATION
A form of heat treatment that kills certain vegetative bacteria and/or spoilage organisms in milk and other foods. Temperatures below 100°C are used.

PATHOGEN
A biological agent (such as a virus or bacterium) which can cause disease.

PREMIXTURE or PREMIX
A mixture of additives, or a mixture of one or more additives with substances used as carriers, intended for the manufacture of feedingstuffs (as defined in the Feeding Stuffs Regulations 1991).

PrP (PRION PROTEIN)
A normal protein that becomes protease resistant in infected tissue and accumulates around lesions in the central nervous system in some spongiform encephalopathies. It can also be detected in trace amounts in some other tissues.

PROPHYLACTIC
(Any medicine) used to prevent disease.

PROTEIN CONCENTRATE
A compound feed with a high protein content used either as an ingredient in compound feed production or as a complementary feedingstuff.

RAW MATERIALS
Products, whether or not containing additives, which can be marketed as straight feedingstuffs, or for the preparation of compound feedingstuffs or as carriers of premixtures.

RENDERERS/RENDERING
See Annex 2.4.

SILAGE AGENTS
Substances added to the silage crop to improve the ensiling process by, for example, improving the fermentation process or reducing the formation of effluents.

SOLVENT EXTRACTION
See Annex 2.4.

STRAIGHT FEEDINGSTUFF
A vegetable or animal product in its natural state, fresh or preserved, and any product derived from the industrial processing thereof, and any single organic or inorganic substance, whether or not it contains any additive, intended as such for oral animal feeding (as defined in the Feeding Stuffs Regulations 1991).

SULPHONAMIDE
Synthetic antibacterial compounds which contain in their chemical structure the nucleus 4-aminobenzenesulphonamide. Examples of the group include sulphadimidine, sulphamethoxazole, and sulphadiazine.

TALLOW
Animal fat obtained through the rendering process (see Annex 2.4).

THERAPEUTIC
(Any medicine) used to treat disease.

TITRE
Concentration (eg of antibody in serum).

TOP DRESSING
The administration of a medicine by sprinkling it on to normal feed, by mixing it with a small quantity of normal feed, or by feeding a medicated feed concentrate to cattle by raking it into the top few inches of silage.

UNDESIRABLE SUBSTANCES
Substances such as heavy metals, pesticides, aflatoxin, and certain plant materials which may contaminate animal feedingstuffs and endanger animal health, or, because of their presence in livestock products, human health.

VERTICAL TRANSMISSION
Transmission of disease from either parent to the offspring, at the time of fertilisation or *in utero*.

WITHDRAWAL PERIOD
The period specified in a current product licence for a veterinary medicine (or in the absence of any such specification, specified in a prescription or veterinary written direction) which is required to elapse between the cessation of the administration of the medicine to the animal, and its slaughter for human consumption, or the taking of products derived from it for human consumption.

ZOONOSIS
A disease common to man and animals, usually in which animals are the main reservoir of infection.

APPENDIX III

ABBREVIATIONS

ADAS
Agricultural Development Advisory Service

AFRC
Agricultural and Food Research Council

BS
British Standard

BSE
Bovine spongiform encephalopathy

CAP
Common Agricultural Policy

CDSC
Communicable Disease Surveillance Centre

CJD
Creutzfeldt Jakob Disease

CVL
Central Veterinary Laboratory

CVMP
Committee for Veterinary Medicinal Products

CVO
Chief Veterinary Officer

CWD
Chronic wasting disease

DANI
Department of Agriculture for Northern Ireland

DH
Department of Health

DHSS(NI)
Department of Health and Social Services (Northern Ireland)

DMRQC
Division of Microbiological Reagents and Quality Control

DNA
Deoxyribonucleic acid

EC
European Community

ELISA
Enzyme-linked immunosorbent assay

FSE
Feline spongiform encephalopathy

HACCP
Hazard analysis critical control point

HSE
Health and Safety Executive

IDCAFS
Interdepartmental Committee on Animal Feedingstuffs

LGC
Laboratory of the Government Chemist

LACOTS
Local Authorities Coordinating Body on Trading Standards

MAFF
Ministry of Agriculture Fisheries and Food

MRC
Medical Research Council

MRLs
Maximum residue limits

OPCS
Office of Population Censuses and Surveys

PAPO
Processed Animal Protein Order

PHLS
Public Health Laboratory Service

PML
Pharmacy and Merchants List

POM
Prescription Only Medicine

PrP
Prion protein

PT
Phage type

RSC
Royal Society of Chemistry

RPSGB
Royal Pharmaceutical Society of Great Britain

SE
Spongiform encephalopathy

SVS
State Veterinary Service

TME
Transmissible mink encephalopathy

VMD
Veterinary Medicines Directorate

VPC
Veterinary Products Committee

WPVR
Working Party on Veterinary Residues in Animal Products

APPENDIX IV

ORGANISATIONS AND INDIVIDUALS WHO SUBMITTED COMMENTS TO THE EXPERT GROUP

Agricultural and Food Research Council
Agricultural Waste Processors' Association
Albright & Wilson Ltd
Animal Health Distributors' Association
Association of Port Health Authorities
Association of Public Analysts
British Association of Feed Supplement Manufacturers Ltd
British Poultry Federation
British Retailers' Association
British Standards Institute
British Veterinary Association
Consumers' Association
Convention of Scottish Local Authorities
Dalgety Agriculture Ltd
B G Davies
Farm & Food Society
Federation of Fresh Meat Wholesalers
Federation of Milk Marketing Boards
Grain and Feed Trade Association
Halib Foods International Ltd
L G Hanson
Institute of Food Science & Technology
Institute of Food Technologists
Laboratory of the Government Chemist
Professor R Lacey
Leatherhead Food Research Association
Local Authorities Coordinating Body on Trading Standards
Meat and Livestock Commission
Midland Trading Standards Liaison Group, Quality Standards Sub-Group
Milk Marketing Board
National Consumer Council
National Council of Women of Great Britain
National Farmers' Union
National Farmers' Union of Scotland
Northern Ireland Grain Trade Association Ltd
Northern Ireland Poultry Federation
Pet Food Manufacturers' Association
Preseli Pembrokeshire District Council
Royal Institute for Public Health & Hygiene
Royal Pharmaceutical Society of Great Britain
Royal Society of Chemistry
Salmon and Trout Association
Scottish Milk Marketing Board
Scottish Agricultural College

Scottish Salmon Growers' Association
Shetland Salmon Farmers' Association
Shropshire County Council Trading Standards
Society For Applied Bacteriology
Society of Feed Technologists
Somerset County Council Scientific Services
Ulster Farmers' Union
United Kingdom Agricultural Supply Trade Association Ltd
United Kingdom Association of Fish Meal Manufacturers
United Kingdom Renderers' Association Ltd
University of Edinburgh Institute of Ecology & Resource Management
University of Surrey School of Biological Sciences
Women's Farming Union

APPENDIX V

ORGANISATIONS AND INDIVIDUALS WHO ATTENDED MEETINGS OF THE EXPERT GROUP

COMMITTEES

Spongiform Encephalopathy Advisory Committee

Dr D A J Tyrrell — Chairman
Dr R Kimberlin
Dr W Watson
Mr D Pepper
Dr H Pickles (Observer)
Mr R Bradley (Observer)
Secretariat: Mr R Lowson (MAFF), Mr T W S Murray (DH)

OFFICIALS FROM GOVERNMENT DEPARTMENTS AND PUBLIC BODIES (UK)

Department of Agriculture for Northern Ireland

Mr G McCracken	Food Policy and Commodity Group Co-Ordination (until March 1992)

Department of Health

Mrs A McDonald	Medical Toxicology Section
Mr T W S Murray	Health Aspects of Environment & Food Division; Joint Secretary of Spongiform Encephalopathy Advisory Committee
Dr L Robinson	Medical Division — Food Safety

Ministry of Agriculture Fisheries and Food

Mr S Bailey	National Analytical Chemistry Specialist, Agricultural Development and Advisory Service (ADAS) (until December 1991)
Dr R Cawthorne	Animal Health (Zoonoses) Division
Mr T Franck	Chemical Safety of Food Division, Secretary to IDCAFS
Mr J Howard	Animal Health (Zoonoses) Division
Mr W Knock	Chemical Safety of Food Division
Mr A Lawrence	Animal Health (Disease Control) Division; former Joint Secretary of Southwood Committee
Dr D Lees	Food Science Division 1 and member of Interdepartmental Committee on Animal Feedingstuffs (IDCAFS) (until January 1992)
Mr R Lowson	Animal Health (Disease Control) Division; Joint Secretary of Spongiform Encephalopathy Advisory Committee
Mr K Meldrum	Chief Veterinary Officer
Mr M Stranks	Senior Professional Officer Nutrition Chemistry, ADAS; Chairman of IDCAFS (until March 1992)

Public Health Laboratory Service

Dr C L R Bartlett	Director, Communicable Disease Surveillance Centre
Dr J Cowden	Communicable Disease Surveillance Centre
Dr S Hall	Field Services & Training Division Communicable Disease Surveillance Centre
Dr P Hamilton	Division of Biologics, Centre for Applied Microbiology & Research
Dr T Humphrey	Food Research Laboratory, Church Lane, Exeter
Dr D Roberts	Central Public Health Laboratory
Sir Joseph Smith	Director
Dr J Threlfall	Division of Enteric Pathogens, Central Public Health Laboratory
Mrs L Ward	Division of Enteric Pathogens, Central Public Health Laboratory

Laboratory of the Government Chemist

Dr N T Crosby	Head of Fertilisers and Feedingstuffs Section

Royal Pharmaceutical Society of Great Britain

Mr D Thomas	Head of Animal Medicines Department
Mrs J Wingfield	Senior Administrator, Law Department (until September 1991)

Veterinary Medicines Directorate

Mr C Bean	Administrative Head of Feed Additives Branch
Mr J O'Brien	Professional Head of Feed Additives Team

Central Veterinary Laboratory

Mr R Bradley	Head of Pathology Department (Observer on Spongiform Encephalopathy Advisory Committee)
Dr A Cullen	Head of Poultry Department (until November 1991)
Dr C Wray	Bacteriology Department

Local Authorities Coordinating Body on Trading Standards

Mr R Andrews	Kent County Analyst
Mr N Edwards	Fife, Trading Standards Officer
Mr D Roberts	County Trading Standards Officer, Shropshire
Mr A Williams	County Trading Standards Officer, Gloucester (until May 1992)

TRADE ORGANISATIONS

National Farmers' Union

Mr B Gill	Vice President
Mr R Parsons	Member of Commercial Services & Transport Committee
Ms A Peterson	Feedingstuffs Advisor

United Kingdom Agricultural Supply Trade Association

Dr B Cooke	Chairman of Scientific Committee
Miss J Nelson	Feed Manager
Mr H I Smith	Chairman of Feed Executive Committee
Dr D Williams	Chairman of Legislation Sub-Committee

APPENDIX VI

VISITS TO ESTABLISHMENTS UNDERTAKEN BY EXPERT GROUP MEMBERS

United Kingdom

BOCM Silcock Laboratory
Barlby Rd
Selby, North Yorkshire

Mr M J Cowan — Chairman
Mr A W Davie — Operations Director
Mr D J Loane — Feeds Director
Dr D R Williams — Chief Chemist & Company Quality Manager

Prosper de Mulder Ltd
Ings Road, Doncaster, South Yorkshire

Mr A J de Mulder — Managing Director
Dr S Wilson — Group Product Manager

Dunstaffnage Marine Laboratory
Oban,
Argyll

Professor J B L Matthews — Director
Dr Bullock — Assistant Director

Golden Sea Produce Ltd
South Shian, Connel, Argyll

Dr A Browne — Director

BP Nutrition (UK) Ltd
Wincham, Northwich
Cheshire

Mr R Watret — Fish Feed Expert

Massey Brothers (Cranage) Ltd,
Cranage Mill, Cranage, Holmes Chapel
Cheshire

Mr Richard Massey — Deputy Director

Belgium

EC Commission Officials, Directorate General VI (Agriculture), Directorate BII (Quality and Health)

Division 1 — Legislation relating to crop products and animal nutrition

Mr G Hudson — Head
Mr J Gaster
Mrs J Marsden
Mr J Thibeaux
Mr R Valls Pursals

Division 2 — Veterinary and Zootechnical Legislation

Mr B Hogben
Mr A Lawrence
Mr B Marchant
Dr O Rohte

Netherlands

Dr J J Bakker   Ministry of Agriculture and Fisheries
Mr Johan den Hartog   Commodity Board for Feedingstuffs, Secretary

Norway

Royal Norwegian Ministry of Agriculture

Department of Veterinary Services

Mr P Folkestad   Deputy Director, Head of Aquaculture Section

Aquaculture Section

Mr N Baalsrud   Superintending Veterinary Officer

Animal Health and Welfare Section

Ms M Kristiansen   Executive Officer
Mr T Solbakken   Superintending Veterinary Officer, EC Affairs

The National Agriculture Inspection Service

Mr K Flatlandsmo   Head of Section
Mr H Birger Glende   Senior Executive Officer
Mr A Froslie   Chairman of Advisory Board for Ruminant, Pig and Poultry
                       Feedingstuffs (Expert on Veterinary Toxicology and Nutrition)
Mr H Ingar Myhre   Senior Executive Officer
Mr J Opstvedt   Chairman of Advisory Board for Fishfarm Feedingstuffs (Expert
                       on Fish Nutrition)

Printed in the United Kingdom for HMSO   Dd294741   C9   7/92   17647